SCHRÖDINGER'S RABBITS

SCHRÖDINGER'S RABBITS

the many worlds of quantum

Colin Bruce

Joseph Henry Press
Washington, DC

Joseph Henry Press • 500 Fifth Street, N.W. • Washington, D.C. 20001

The Joseph Henry Press, an imprint of the National Academies Press, was created with the goal of making books on science, technology, and health more widely available to professionals and the public. Joseph Henry was one of the founders of the National Academy of Sciences and a leader in early American science.

ISBN 0-309-09051-2

Cover design by Michele de la Menardiere.

Hand-drawn illustrations by Laura Dawes from sketches by Colin Bruce.

Printed in the United States of America

Dedicated to

Paul Dirac
physicist extraordinary
who believed we must seek visualizable processes

and

Jim Cushing
philosopher of science
who believed we must find local stories

PREFACE

Does the weirdness of quantum indicate that there is a deep problem with the theory? Some of the greatest minds in physics, including Einstein, have felt that it does. Others prefer to believe that any conceptual difficulties can be ignored or finessed away. I would put the choice differently. The flip side of a problem is an opportunity, and the problems with the old interpretations of quantum present us with valuable opportunities.

First, there is the hope of finding ways to think more clearly about the subject. I have several times seen highly respected scientists—physicists whose ability to work with the math of quantum mechanics is certainly better than my own—make appalling freshman howlers in describing what the result of an experiment would be, because their qualitative thinking about such matters as quantum collapse was as fuzzy as everyone else's. Better conceptual tools are badly needed—and now they are becoming available.

Second, there is the possibility that a clearer view of quantum will cause us to see the universe in a fundamentally different way, with implications both practical and philosophical. Then, as has happened so many times in physics, the resolution of a seemingly arcane problem will open our eyes to great new wonders. To ignore such an opportunity would be sheer cowardice.

The past few years have seen a sudden explosion of light in the murkier corners of quantum. The old stories, involving such quaint characters as dead-alive cats and conscious observers with the power to "collapse" the whole universe, or even split it in two, are passé. There are new stories to choose from, one of them particularly promising. It restores us to a classical universe where things behave predictably rather than randomly and where interactions between things are local rather than long range. But it comes at a price. We must accept that the universe we inhabit is much vaster than we thought, in an unexpected way.

Although the many-worlds view was invented in the United States, it is in Europe, and especially in Oxford, that it has developed to maturity. That is my good luck, for I have had the privilege of seeing the process at first hand. Here I describe the remarkable new picture that has recently emerged, which I dub the Oxford Interpretation.

My warmest thanks go to my editor Jeff Robbins at Joseph Henry Press for his vision and determination in ensuring that this book came to be. Also to many physicists and philosophers at Oxford and elsewhere for valuable advice and discussion, including in particular Harvey Brown, David Deutsch, Roger Penrose, Simon Saunders, David Wallace and Anton Zeilinger. Special thanks to Lev Vaidman, Jacob Foster, and Heather Bradshaw, who read the manuscript at an advanced stage and made many useful comments. Responsibility for any mistakes that remain, and any controversial opinions expressed herein, is of course entirely my own.

Colin Bruce
Oxford, 2004

CONTENTS

CHAPTER 1

A MAGICAL UNIVERSE

As a teenager, I was a great fan of science fiction and fantasy. The stories I most enjoyed were those set in a universe very like our own, but with an extra twist—some magical feature that made it much more fun to live in than the mundane world I knew. Then I grew up and discovered something wonderful. Our own real universe does in fact contain at least one magical feature, a built-in conjuring trick that seems to violate all the normal rules. Here is a demonstration.

Imagine that a conjurer of impressive reputation is in town and one night you go along to his show.

"For my next trick," he says, "I want a couple from the audience." To your embarrassment he points straight at you and moments later you find yourself on stage with your partner.

"I would like to give you a chance to get rich," he says, pointing to a large pile of scratch-off lottery cards, all seemingly identical, and looking like the one in Figure 1-1.

"All you have to do to win a prize," he goes on, "is select one of these cards, and tear it in half between you. Each take your half of the card and scratch off 1 of the 60 silvered spots on the clock face to

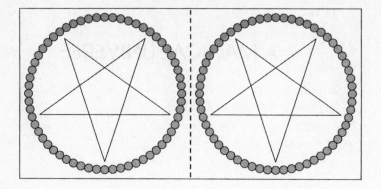

FIGURE 1-1 Lottery card.

reveal the color, either black or white. If the spots you scratch turn out to be different colors, you win $500. And it costs only $10 to play!

"Of course each of you is allowed to scratch off only one spot on your respective half of the card. And there is one further rule: To win the prize, you and your partner must choose spots exactly *one place apart* on the clock face. For example, here is a card that won for two lucky, lucky people on yesterday's show." He shows you and the rest of the audience the card shown in Figure 1-2.

"You must allow me some secrets, so I will not tell you exactly how the cards are colored. But I will tell you this much. Half of all the

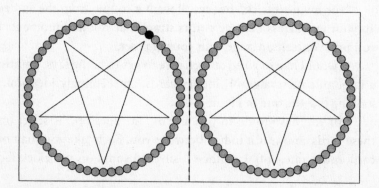

FIGURE 1-2 Winning lottery card.

spots are black, and half white. Also if you and your partner were to scratch off the *same* spot on each clock face, you would always get the same color—both spots would be black, or both white. But if you were to scratch off spots exactly 90 degrees apart from each other, you would always get opposite colors; white and black, or black and white."

It seems like a bargain, but you hesitate. How do you know he is telling the truth? "I'm from this town, and you've got to show me," you reply, to cheers from the rest of the audience. The conjuror nods, unsurprised.

"Be my guest," he says. "You and your partner may choose any card from the pile, and perform either of those two tests—scratch the same spot on each half, or spots 90 degrees apart on each half. Do that as many times as you like. If you prove me a liar, I'll pack up my magic show and take an honest job!"

You and your partner duly pull out and test numerous cards. The results confirm the conjurer's predictions, as shown in Figure 1-3a and b.

Is it worth playing the game? You think carefully. First, the left and right halves of each card must be identically colored—otherwise you would not be sure of getting the same color every time you scratch spots in matching positions. Second, there must be at least one place in each 90-degree arc where the color changes between black and white. If any card had an arc of more than 90 degrees all one color, you could sometimes scratch spots 90 degrees apart and get the same color.

The most obvious guess—and no doubt what the conjurer intends you to think—is that the cards are colored in four quarters, as shown in Figure 1-4a. There cannot be *fewer* segments, as shown in Figure 1-4b, because then you could scratch spots 90 degrees apart and get the same color, which never happens. They might be divided into *more* segments, as shown in Figure 1-4c, but that would actually increase your chances of winning—there are more black-white boundaries to hit.

As you go round the circle, from spot to spot, you take a total of 60 steps. At least 4 of those steps—maybe more, but certainly no fewer—involve a color change, stepping from a black spot to a white one or vice versa. It follows that the chance of a color change on any particu-

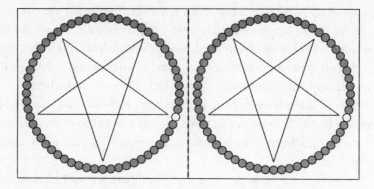

FIGURE 1-3a Corresponding spots scratched: colors always the same.

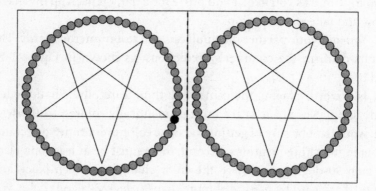

FIGURE 1-3b Spots 90 degrees apart scratched: colors always opposite.

lar step is at least 1 in 15. At those odds, it is certainly worth risking $10 to win $500, and you accept the bet and select a card. The conjurer beams.

"To make the game a little more dramatic, I will ask you to tear the card in two between you, and each take your half into one of the curtained booths at the back of the stage." He points to two curtained cubicles rather like photo booths. "Each of you should scratch a spot of your choice, then stand and hold the card above your head. After a few seconds the curtains will be whisked away, and you and the audience will see immediately whether you have won. Of course, you can

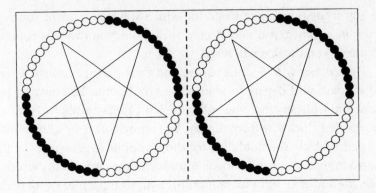

FIGURE 1-4a Could the cards be printed in this pattern, alternating quarters black and white?

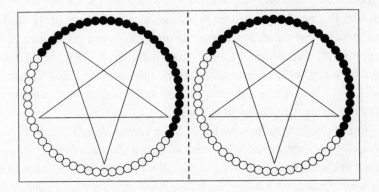

FIGURE 1-4b Or this pattern, alternating halves black and white?

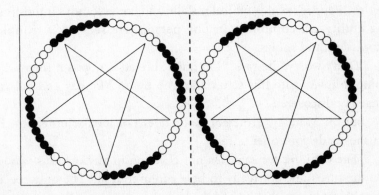

FIGURE 1-4c Or this pattern, many alternating black and white segments?

use any strategy you like to decide which spots to scratch. You may confer in advance, you may decide at random, you can toss coins or roll dice if you think it will help."

He watches with a smile as you and your partner choose a card, tear it apart, and depart to your respective booths. You have in fact decided in whispers that you will scratch off spots number 17 and 18, as measured clockwise from the top. You scratch off your spot and it is revealed as black. You hold the card above your head as instructed. But when a moment later a drumroll sounds and the curtains are whisked aside, the audience sighs in disappointment; your partner's spot is also black. You have lost the game.

As you take your seats again, you are not particularly surprised or disappointed. After all, you reckoned you had only 1 chance in 15 of winning. But now the conjurer proceeds to call up more of the audience, two by two, and put them through the same procedure, 100 couples in all. Out of the lot, only one couple wins—you would have expected six or seven. The winning odds appear to be 1 in 100 rather than 1 in 15, and the conjurer has made a tidy profit. There seems to have been some mistake in your logic.

You are feeling quite worried. If your reasoning can mislead you this badly, you are obviously at risk of being cheated right, left, and center. As the crowd flocks toward the exits at the end of the show, you are therefore delighted to see your longstanding friend and colleague, Emeritus Professor Cope. Professor Cope might be old, but he is the most impressive guy you know. This man has Einstein's scientific intuition, Popper's philosophical insight, and James Randi's fraud-busting ability, all combined in one person. He sees your troubled expression, and smiles.

"Don't worry," he says. "I'm quite sure all is not as it seems. I'm going to investigate this setup. I'll drop by on Monday and tell you what I've discovered."

But on Monday, Professor Cope does not look triumphant. He brushes aside your offer of tea.

"The conjurer we saw was not cheating in any obvious way. In fact, he turns out not really to be a conjurer at all. The only special thing about him is that he had the luck to come across the supplier of these extraordinary cards. I managed to track down this supplier, and ordered a big batch for myself. I've been testing them under controlled conditions, and the results are still exactly the same as you saw at the show the other night."

Your mouth falls open. "But how can that be?" you ask.

Professor Cope smiles. "To quote a respected source, 'When you have ruled out the impossible, what remains, however improbable, must be the truth.' The only way to get the results we see is if the two cards contain some internal mechanism that changes the spot color depending on circumstances. For there is no fixed coloring that can explain the results.

"But the card halves must also be in some kind of radio contact with one another. If they operated independently, there is no way the colors could then always match when you scratch the same place on each. One card half on its own could not tell whether the other half had that same spot scratched, or a different one.

"So the two halves must be in communication. Each half somehow knows which spot was scratched on the other, hence the angle between the two spots, and the color revealed on each card is selected accordingly. It is amazing even in these days of advanced electronic technology, but each card must include something like a miniaturized radio transmitter and inks that can change color. I am going to prove my hypothesis by separating the two halves of a card in such a way that communication between them is impossible. Then we will see the mysterious correlation between the two parts vanish. I will tell you the result next week."

But the following Monday, Professor Cope does not look any happier.

"I tried testing halves of the lottery cards in lead-lined cellars several miles apart, and still got the same disconcerting results. So I borrowed two of those special security cabins-on-stilts used by the military and diplomats for top-secret conferences inside embassies. They are designed to allow absolutely no signal of any kind to leak out. Yet when lottery cards were scratched inside each of them, the results were still the same.

"Then I had a better idea. It occurred to me that there is no such thing as a *perfect* shield for radio and other waves. So I tore a big batch of cards in half, and mailed one set of halves to Australia. I also built a mechanism that allowed a card to be scratched, and the color revealed to be permanently recorded at an exactly timed instant. The whole process takes only a fraction of a second. I had my colleague in Australia build a similar apparatus.

"We proceeded to scratch cards here and in Australia at exactly synchronized moments. Now according to Einstein's theory of relativity, nothing can travel faster than light—neither matter nor radiation of any kind. As many popular accounts have described, if you could send a signal faster than light, you could also send one backward in time.

"The distance from here to my colleague's laboratory in Sydney, even if you take a shortcut through the center of the Earth, is nearly 8,000 miles. It takes light about a 20th of a second to make the journey, a time just perceptible to human senses. My automatic card-scratching-and-color-measuring apparatus works much faster than that. So there was absolutely no way that either the card here could send a signal to its twin in Australia, or the Australian card could send a signal here, before both cards had to decide what color to reveal."

He pulls a whiskey bottle from his pocket and takes a swig. "I would have bet my life's work that under these circumstances, the strange correlations would disappear. But they did not.

"Well, no one is going to call me an intellectual coward. If I have proved the existence of faster-than-light, backward-in-time signaling of unlimited range, so be it. One card half must be sending an instantaneous and undetectable signal to the other. There you have it!"

You shake your head sadly as you see him out. But the following evening, he calls in looking much happier.

"Forget all that nonsense I was talking yesterday about faster-than-light signaling," he says. "After I left you, I spent some time trying to figure out how to harness the cards' instant links to transmit information. It would be handy to be able to talk to an astronaut in distant space without the normal time lag while the radio waves go to and fro, and even better if you could send a message with tomorrow's racing results back in time to yourself! But there is no way to use the cards to do these things, because you have no way to influence the color of the spot you scratch off. It is always 50-50 whether it is black or white. It is only *after* you compare the card with its other half that the strange correlation is revealed.

"I decided that because any supposed faster-than-light signaling mechanism is not available outside the cards' internal workings, Occam's razor—that rule of science that demands that one should always seek the simplest explanation, avoiding unverifiable hypotheses—required me to dispense with it. I now have a better theory.

"The correlations are surprising if you and your partner can make genuinely free or random decisions as to which spots you are going to scratch. But suppose those decisions have in fact been preordained for all time? You feel subjectively that you are freely choosing which spot to scratch, but actually the movement of the electrons that would make your neurons fire in that way was inevitable from the start of the universe—there is no free will. Similarly, if you use a randomizing device like dice or a roulette wheel to help you choose the spots, its motion and outcome were also predictable.

"The lottery cards must have been manufactured by an all-knowing alien who simply knew in advance exactly which spot on each half would be scratched, and printed the cards accordingly. Try as you will, he has foreseen your every move! This might sound startling, but it explains away the apparent paradox."

You do not know what to think as Professor Cope takes his leave. It certainly seems an alarming amount of philosophical baggage to explain a set of trick lottery cards. At six o'clock the next morning the doorbell rings again. You stagger down bleary-eyed in your bathrobe to find a disheveled but triumphant Professor Cope on the doorstep. The whiskey bottle protruding from his pocket is nearly empty.

"I have it," he says happily. "It is amazing how late-night thought, assisted by strong liquor on an empty stomach, can strengthen one's facility for philosophical reasoning. I was worrying about a non-problem! You would agree that science can concern itself only with things that are actually observable, rather than mere hypotheticals?"

"I suppose so," you agree cautiously.

"Good! Now, you are a conscious observer and, as such, the only hard data you are entitled to reason about are the things that you have actually observed. All that precedes observation is mere will-o'-the-wisp, hypothetical, unreal. Let us consider your point of view at the moment you scratch off the lottery card. You see a color, black or white—perfectly reasonable. A little later you see your partner's card, which is also black or white—perfectly reasonable. The only problem comes from your worrying about the hypothetical 'I wonder what my partner's card was?' in advance of actual knowledge, when it was still an open question. Your partner's card wasn't anything until you found out what it was! When it did become something, it conformed to the claimed statistics for the admittedly unusual cards. But there is no problem for physics, as long as you have a formula to calculate the statistics. And no problem for philosophers, as long as you do not ask questions that are in fact meaningless because you are confusing hypotheticals with hard data. So, no problem!"

This is all a bit much at 6 a.m. "But isn't that a bit solipsistic?" you ask. "I mean, what about my partner's point of view? Are you really saying that it was meaningless for her to wonder what color the spot on my card was until she saw it? Confound it, *I* had seen it, and it was black, not hypothetical!"

"Solipsism, schmolipsism," says Professor Cope crossly. "I have explained things from your point of view, the only one you should legitimately be concerned with." And he turns on his heel.

It is sad to have witnessed the decline of a once great mind, but you do not see Professor Cope for some time after that, and gradually you forget about the matter. After all, you have plenty of practical everyday problems to worry about. Then one day, Cope strides confidently up to you in the shopping mall and grasps you by the arm.

"I am sorry about the nonsense I was talking a while back," he says immediately. "I have given up the philosophizing business, and gone back to hard physics. I now have a perfectly consistent explanation for the lottery cards that does not involve dubious philosophical assumptions, backward-in-time signals, or any other rubbish of that kind. Let me buy you lunch. In fact, in a sense I will buy you a *lot* of lunches."

He steers you into a nearby restaurant, and laughs inordinately when the host asks how many in your party. "Just two," he finally gets out, "that is, as far as *you* are concerned, young man." As you start on the soup, he launches into his new story.

"Like all conjuring tricks, it is quite simple when you see how it is done," he says. "The truth is, the maker of the lottery cards had a rather special kind of duplicating machine."

"Well, I suppose it takes something a bit fancier than a standard printing press to make those scratch-off cards—" you say, but break off, because Cope is shaking his head vigorously.

"I am talking about something rather grander than that. Those lottery cards were manufactured by an all-seeing and all-powerful alien who can duplicate multiple versions of the universe at will!

"At the point where two people scratch off spots on the two separated halves of one of his lottery cards, the alien simply multiplies up the numbers of versions of reality to produce statistics that will conform to his rules. Thus if you each scratch off a spot in the same place, he creates two versions of the universe. In one, you and your partner both hold a black spot; in the other you both hold a white. From your point of view—that is to say, from the point of view of any one version of you—the spot color is entirely random and unpredictable, yet you will always find that it is the same as your partner's.

"If you scratch off spots 90 degrees apart, the alien again creates two versions of the universe, but this time in one version you hold a black spot and your partner a white; in the other you hold a white spot and your partner a black. Again, from any individual's viewpoint the color of their spot is unpredictable, but it will always be the opposite of their partner's.

"Now for the clever bit. If you scratch off spots exactly one place apart, the alien creates 200 versions of the universe. In one of those, you hold a black spot and your partner a white. In 99, you and your partner both have black spots. In another 99, you both hold white spots. And in a final one, you hold a white spot and your partner a black. Again, you—or to be more precise in my language, any one *version* of you—experience getting a spot of entirely unpredictable color, but then find that your partner holds the opposite color just 1 percent of the time." He beams proudly. "A beautifully simple idea, is it not?"

But you have already picked up your coat. There are limits to the nonsense you will listen to, even in return for a free lunch. You have decided that the best way to retain your sanity is to try and forget the whole business.

∞∞∞

In real life, we cannot escape the challenge so easily. As many readers will of course have realized, the apparently extraordinary lottery cards are merely behaving in the way that *all* the material in our mundane, everyday world does. Very similar effects can be demonstrated using the simplest particles of which our universe is built, the photon and the electron, the basic units of light and matter. Measuring the spin of an electron, or the polarization of a photon—scratching its lottery card, so to speak—seemingly has an instantaneous effect on the outcome of a measurement of another particle some distance away.

The formal name for this puzzle is the EPR paradox, after its originators Einstein, Podolsky, and Rosen. It is the most puzzling feature of the modern formulation of physics known as quantum theory. For half a century, attempts by physicists and philosophers to explain this behavior have verged on the bizarre. They are only mildly caricatured above. The purpose of this book is to find a more commonsense account of how the conjuring trick is done.

CHAPTER 2

CLINGING TO THE CLASSICAL

What is the real-life manifestation of the problem that has gotten scientists in such a spin? It started relatively innocuously about a century ago, with a new twist in an ancient debate—about whether light was composed of waves or particles.

This question had been considered settled at the end of the 18th century, through an ingenious experiment by the British natural philosopher Thomas Young, which involved passing light through slits. When a wave passes through a narrow slit, it tends to spread out on the other side. You can see this happen when a water wave passes through the gap in a harbor wall. It does not just continue on its original straight-line track, but spreads out so that all the boats in the harbor end up bobbing up and down. Light behaves in just this way when it passes through a narrow slit.

Particles don't generally do the same, but it's easy to envision how they could be made to. Suppose you were rolling bowling balls toward a narrow gap in a fence. It would be easy to place some springy twigs around the gap so that the bowling balls were deflected by random angles as they passed through. Then a stream of bowling balls being rolled toward the gap would spread out over a range of angles on the far side, just as a wave does. It was evident to Young and others that if light consisted of a stream of particles, these might be scattered when passing close to solid matter (as when passing through a narrow slit) by something analogous to the springy-twigs effect. So the spreading is not in itself convincing evidence whether light consists of waves or particles.

However, a cleverer experiment involving two slits appeared definitive. Imagine a blindfolded man rolling bowling balls toward a fence in which either or both of two narrow gates might be open. The gates have springy twigs placed so that any ball passing through a gate is deflected by a random angle; behind the gates is a line of catchment trenches into which the balls fall. It is fairly obvious that the effect of opening both gates is that each trench gets the sum of the balls it would have gotten if only the left gate was open and those it would have gotten if only the right gate was open, as shown in Figure 2-1. Certainly, closing a gate can never increase the number of balls going into a given trench. The bowling balls are of course behaving like particles.

But now suppose we do a similar experiment with waves. For example, we could flood the bowling green and generate water waves of a particular wavelength, as shown in Figure 2-2. As waves strike the barrier at the back, water slops over it (more where the waves are higher, obviously), gradually filling the catchment trenches.

When only one gate is open at a time, the accumulation of water after an appropriate number of waves have been generated is very similar to the result obtained with the bowling balls, as shown at the top of Figure 2-2. But when both gates are opened simultaneously, something quite different happens. Now some trenches that got quite a lot of water when only one gate was open get less, or even none at all.

A little thought reveals why. At points like X, the peak of a wave from one gate always coincides with the trough of a wave from the other. (Peaks are shown as solid lines, troughs as dotted lines.) This leaves the net water depth unchanged at all times, so no water flows over the barrier. The waves from the two gates are said to cancel at such points, and this phenomenon is called *interference*. This is behavior that particles cannot possibly exhibit; opening an extra gate never reduces the quantity of balls reaching any trench. Young realized that this was a neat way to distinguish waves from particles. When he tried the two-slit experiment with light, the results corresponded to Figure 2-2. A pattern of light and dark stripes was visible at the back of the apparatus, and points like X received no light at all. An age-old debate appeared to have been settled; light definitely consisted of waves.

☙☙☙

But more than 100 years later, at the start of the 20th century, this picture was thrown into confusion. By then, it was known that solid matter was composed of the tiny particles the Greeks had hypothesized, called atoms, and moreover that atoms were composed of positively charged central nuclei and negatively charged particles called electrons. Electrons could be detached from their parent atoms and made to flow about within a solid material, as when an electric current flows down a wire, and even sprayed into empty space, as happens inside a TV tube. It had become possible to do experiments that involved light interacting directly with electrons. This is not a history book, so I am going to describe only the most definitive of these experiments, which is now called the Compton effect.

Back in the 1920s, Compton arranged to spray electrons into a vacuum, and then shine a bright light of a particular color onto them at right angles as shown in Figure 2-3. It had long been known that light radiation carries momentum as well as energy, so that light shining on a surface exerts a slight pressure. The pressure is small by ordinary standards; if you hold your cupped hands up to the Sun, the force on your palms is about a millionth of an ounce. Nevertheless, light pressure is strong enough to propel a kind of spacecraft called a solar sail, and certainly strong enough to deflect a beam of lightweight particles like electrons.

If light consisted of waves, it would be reasonable to expect that all the electrons would be deflected by a similar amount, as on the left of Figure 2-3. But what really happens is quite different, as shown on the right. Most of the electrons are completely unaffected. But an occasional electron is deflected by a large angle. This is characteristic of two streams of particles intersecting. Think of the electrons as cannon shells and the photons as lighter but faster machine-gun bullets. If a cannon shell happens to be hit by a bullet, it is deflected quite sharply,

FIGURE 2-1 Blindfolded bowler with one gate open (top) and two gates open (bottom). Balls that hit the fence are assumed to be removed; the pattern shown is the average that would result if the experiment was repeated a large number of times.

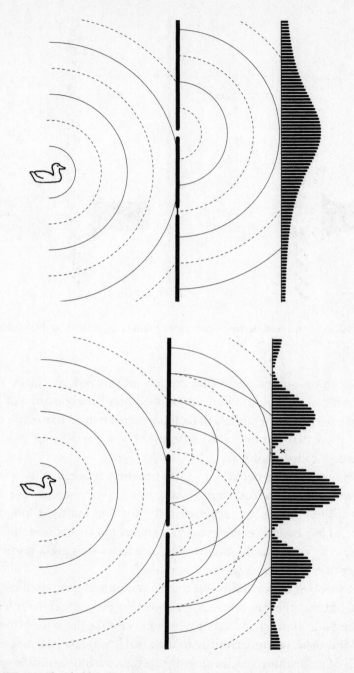

FIGURE 2-2 Flooded bowling green with one gate open (top) and two gates open (bottom).

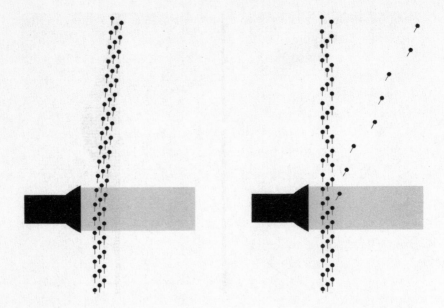

FIGURE 2-3 Stream of electrons intersects a beam of light: two possible outcomes.

but all those cannon shells that are not hit proceed on exactly their original course. Compton's result implied that light consisted of bullet-like particles. If a particle of light happened to hit a particular electron, then that electron was deflected. These particles of light are nowadays called photons.[1]

How could this be? When light is traveling, it behaves like a wave, spreading out to explore every possible route open to it as a wave does, even if these routes are centimeters (or, for that matter, kilometers) apart, as in a two-slit experiment. But when light strikes something, it appears at very specific points, like hailstones striking a pavement rather than floodwater washing across it.

One obvious possibility was that light is indeed composed of photons, but the photons are so numerous that they somehow interact, jostling one another so as to give rise to wavelike behavior. After all, the kind of wave most familiar to us, a water wave, is just the visible result of many tiny particles moving together, pushing against one another as they do so. Just as atoms are very small physical things, photons are very tiny packets of energy. A lightbulb emits about 10^{20} (that stands for one followed by 20 zeros, 100 billion billion) photons of visible light every second. This is roughly the same as the number of atoms in 1 cubic millimeter of solid matter. Perhaps just as billions of air molecules jostling one another can produce a sound wave, and billions of water molecules jostling one another can create a geometrically perfect ripple on the surface of a liquid, billions of photons jostling one another could produce light's wavelike action?

Nobody was very happy with this picture, though. The problem is that there are not really enough photons around to produce wavelike interactions. That might sound paradoxical—10^{20} is a huge number—but let's do some figuring. Photons travel so fast that a photon emitted from a lightbulb in an ordinary room has a lifetime of only a few billionths of a second before it hits something or escapes through a window, meaning that there are some 10^{12} photons present in the room at any time. That's a density of only about 10 photons per cubic millimeter, compared to 10^{16} air molecules per cubic millimeter.

Another way to look at it is that if we put a soap bubble with a radius of 1 meter and a thickness of 1 wavelength of visible light around the bulb, its skin would contain only 100,000 photons at any instant—only 1 per square centimeter. Yet if photons really were particles, they would have to be tiny things. An appropriate unit of measure to use here is the Angstrom, 1 ten-billionth of a meter. The atoms in a typical solid are 2 or 3 Angstroms apart. When a photon hits a solid, it usually interacts with just a single atom. A particlelike photon would therefore presumably be, at most, 1 Angstrom in diameter. Could such a tiny thing really jostle other corpuscles millimeters or even centimeters away from it? The problem gets worse when you take into account that, even with naked-eye observation, light's wavelike behavior can be seen at illumination levels thousands of times less than a brightly lit room, when individual photons are centimeters or even meters apart.

In fact, photon jostling can be ruled out altogether. With slightly more modern technology than Young's, we can lower the level of illumination inside a two-slit apparatus to the point where there can only be a *single* photon in it at any given time, and place sensitive photographic film at the back. We leave the experiment to run for a while, then develop the film. The pattern of light and dark stripes is still visible on the film. Somehow each and every photon, a thing so tiny that it interacts with just one atom when it strikes a solid surface, has had its trajectory influenced by the presence and position of both slits. How could each photon possibly have explored, or somehow been aware of, both possible routes? Figure 2-4 shows the contrasting pictures of light as consisting of waves on the one hand, and photons on the other. The left picture shows light as it typically behaves in flight, the right as it typically behaves when it hits something.

Many textbooks describe this as behavior that cannot be explained in terms of any classical picture, a picture in which some kind of behind-the-scenes machinery does definite things at definite locations and times. But that is an oversimplification. Let us demonstrate a determination that is going to guide us throughout this book. We are going to stick stubbornly to the notion that we will explain what is going on in a commonsense, visualizable way. There *is* such a way to explain the behavior of light going through a two-slit apparatus, and Einstein, among others, was fond of it.

The concept is called pilot waves. Suppose that any light source

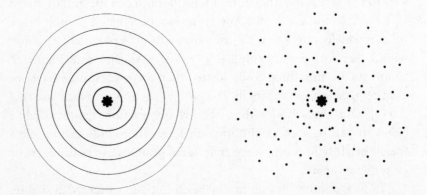

FIGURE 2-4 Two contrasting pictures of light from a point source. Is it emitted as concentric waves, like the ripples from a fisherman's float bobbing up and down in the water, or as individual photons flung off in random directions like sparks from a firework?

actually emits two kinds of thing. The first are waves as shown on the left of Figure 2-4; however, the waves themselves are completely invisible and imperceptible to us. But the light source also emits photons, as shown on the right. The trajectories of the photons are guided by their interactions with the invisible waves.

Let us return to the bowling-green picture of Figures 2-1 and 2-2. Suppose we flood the bowling green as in Figure 2-2—but now throw a bowling ball into the water. The ball's motion generates a gentle wave, and the ball travels along with the wave, being guided by it. The bowling ball can obviously go through only one of the gaps in the fence, but the wave goes through both, and continues to guide the ball to its final destination. Although the bowling ball is always in one place, the wave has explored both possible routes, and a pattern like that in Figure 2-2, but with the trenches now full of bowling balls rather than water, can arise quite naturally. We have solved the wave-particle paradox! (We'll assume that the bowling balls are light enough to float. You might like to think of the ball as a surfer riding a wave, who prefers to be at the highest point of the wave. He is not perfectly successful, but is most likely to be found where the wave is highest, least likely where it is lowest.) [2]

∞∞∞

As Compton experimented further with his electron-deflecting apparatus, he confirmed another property of photons. Increasing the intensity of the light increased the number of electrons knocked aside, but not the amount by which each electron was deflected. The greater intensity increased the number of photon-particles, but not the amount of momentum carried by each. On the other hand, changing the color of the light *did* change the angle by which each electron was deflected. Blue photons knock electrons aside at almost twice the angle that red photons do, indicating that each blue photon carries twice as much momentum or "punch" as a red one.

It had long been known that the color of light is really just the way we perceive its wavelength. For example, blue light has a wavelength of approximately 4,000 Angstroms, and red light approximately 7,000 Angstroms. Compton's result verified that the momentum of individual photons is related to the wavelength of the light involved—the shorter the wavelength, the more the momentum and energy carried by each individual bullet of light. The actual formula is this:

$$\text{Wavelength} = 6.62 \times 10^{-34}/\text{Momentum}$$

(The quantity 6.62×10^{-34} stands for 6.62 divided by the number 1 with 34 zeros written after it, that is, .000000000000000000000000000000000662. This quantity appears in many equations of modern physics, and is known as Planck's constant.)

This leads to a curious thought. Why should this formula apply only to particles of light, and not to particles of matter as well? If it does apply to solid objects, then the wavelength associated with large things like bowling balls will be incredibly tiny. But the wavelength associated with minute things, like atoms when they are moving slowly, will be quite large. It turns out that when we repeat the bowling-ball experiment of Figure 2-1 on a small enough scale, using individual atoms as the balls, then the results are again like those of Figure 2-2. An atom that can sometimes get to X when one gate is open cannot do so when both gates are open! Just as the waves of light can also behave as discrete particles, so the discrete particles of solid matter can also behave as if they were waves.

Once confirmed, the wavelike behavior of matter solved some tough problems that had confronted the early atomic theorists. An early model of the atom—still seen in pictures today—resembled a tiny solar system, with electrons circling the central nucleus like planets circling the Sun. But whereas real solar systems are all slightly different from one another, atoms of the same type all behave in exactly the same way. Take the most basic atom, hydrogen, a single electron circling a single proton. Rather than orbiting the proton at any arbitrary distance, as a planet could, the electron can occupy only certain orbits or energy levels. When the electron switches between two orbits, the amount of energy emitted is therefore always one of a few exactly predictable quantities. This cannot be explained by a purely particle-like electron. If the electron has a wave associated with it, however, then the math predicts that only certain wavelengths will be stable, and therefore describe allowed orbits for the electron, just as a bell can vibrate stably only at certain frequencies corresponding to its harmonics.

This triumph, explaining the quantization of atomic energy levels, is what gives quantum theory its name. But I would like to stress that this wavelike behavior does not apply just to tiny objects like atoms and molecules, but to objects at any scale. To illustrate, I am tempted to ask you to imagine a wall with two slits in it, and a gun capable of firing a cat toward the arrangement, but cats (even hypothetical ones) have already suffered enough in the cause of quantum physics, and Stephen Hawking has threatened to shoot people who mention Schrödinger's cat to him, so I will choose an alternative. I have visited a Rolls-Royce factory where they test their jet engines' ability to survive bird impacts. The apparatus they use is a kind of catapult that fires oven-ready chickens (an accurate model for the largest kind of birds that an aircraft is likely to hit, and available in a range of sizes at the local supermarket) at random angles toward an engine on a test rig. Suppose we remove the jet engine and replace it with a brick wall with two slits in it. Every time a chicken gets through to the far side of the hangar beyond the wall, we make a chalk mark at that point. Eventually we would expect to see a pattern like that of Figure 2-2. With chickens, the scale of the pattern would be incredibly fine, far too fine to measure practicably, but it would be there.

With lightweight particles like electrons, however, the experiment can easily be done. If the experiment shown in Figure 2-1 is done with a source of electrons of appropriate momentum, and hence wavelength (which works out to be electrons traveling at about 1 mile per second, a rather modest speed for an electron), we get an interference pattern as shown in Figure 2-2, at exactly the same scale as one produced by visible light. While they are flying through free space, electrons behave like spread-out waves. Only when they hit something do they remanifest themselves as pointlike objects. Yet we know from other experiments that electrons are much tinier even than atoms. In fact, they are perfectly pointlike insofar as anyone has ever been able to detect. How can this be?

By now I am sure there is an answer on the tip of your tongue—pilot waves! Every time you let fly with an electron (or for that matter with an oven-ready chicken) the action also generates an invisible wave, which guides the subtle motion of the object. This possibility was taken seriously by many physicists at one time, and still is by a few. But guide waves for solid objects raise conceptual difficulties that are not present (or at least not so apparent) when photons are involved.

In the case of a photon, the point where the guide wave comes into existence is well defined. It is created together with its photon when radiant energy is emitted, and effectively dies (or at least ceases to have significant effects on the rest of the universe) when that photon is absorbed. The photon then momentarily appears at one definite point in space—following the period of travel on the guide wave when its whereabouts were unknown—and expires, donating its energy at that particular point. The image of a hapless surfer finally splatted against a harbor wall is unavoidable. After that, of course, it does not matter what happens to the pilot wave. Its only discernible effect ever was to guide the photon; once the photon is gone, you can think of it as passing on to infinity without any subsequent effect on the rest of the universe.

Particles like protons and electrons, by contrast, have very long lifetimes, typically comparable to the age of the universe, during which their initial guide waves presumably continue to exist, spreading farther and farther throughout space. But we do not need to destroy an electron or a proton in order for it to turn up in some definite place during that time.

What causes a particle like an electron to become localized, and appear in one place rather than another? The theoretical answer to that question is deep and problematic. But the immediate empirical answer could not be more straightforward. The electron's location becomes definite when an experimenter measures it! Until such a measurement is made, the electron could be anywhere on its guide wave; afterward, its location can be known (at least temporarily) to an arbitrarily high degree of precision. This sudden localization is a form of what is called quantum collapse.

Such measurement has a curious side effect. It effectively knocks the particle you are measuring off its guide wave. If the blobs in Figure 2-1 represent particles, such as electrons or oven-ready chickens, then any attempt to measure the trajectories of the particles destroys the interference pattern shown in Figure 2-2; instead we again get a result like that in Figure 2-1. It seems that any kind of stuff (whether light or solid matter) can behave either as waves or as particles, but never as both at the same time. If we look at the particles, to try to see which slit they are going through, the wave effects disappear.

At first this sounds like a very strange effect. But what do we really mean when we say that we "look at" the particles? In experimental practice, this translates as: We shine a bright light on them. With normal levels of light, we can see which way an oven-ready chicken is going; with sufficiently bright light, we can even see which way electrons are going. When we do the two-slit experiment with electrons, a perfect interference pattern appears only if the experiment is done in the dark. The brighter the light shone on the electrons, the fainter the interference pattern produced. This washing out of the pattern has nothing to do with whether anyone is watching—be it a so-called conscious observer, a cat, or a camera. We already know that light can affect electrons. There is no reason to assume that anything mystical is going on. It just so happens that the point at which the light becomes bright enough that we can start to tell which way each electron is going is also the point at which the interference pattern starts to disappear.

There is a curious corollary to the wavelike behavior of particles. We find that however bright a light we shine on a small particle like an electron, we can never pin it down perfectly, in the sense of simultaneously knowing its exact position and its exact motion precisely. This, as many readers will recognize, is Heisenberg's famous uncertainty principle in action. But there is a way to explain this, too, in terms of guide waves. A particle can never be completely divorced from a guide wave—in terms of our poetic surfboarder analogy, the surfer always determinedly climbs back on and finds a new wave, however often he is knocked off the old one. Trying to measure the position of the surfer-particle exactly is like trying to squeeze the entire guide wave into a very small space. Much as when the soap in the bathtub tries to escape as you close your hands about it, amplifying the effect of any waves in the tub, so trying to squash a particle's guide wave into a small space tends to induce it to a higher speed.

Just as water waves can make a floating cork bob about a great deal while having no discernible effect on a big ship, Heisenberg's uncertainty principle is much more noticeable with small things, like electrons and atoms, than with large things like bowling balls and cats. In this respect, Heisenberg uncertainty is analogous to the phenomenon called Brownian motion: When small things like pollen grains floating in air are observed under a powerful microscope, they jitter around because the number of air molecules which are at all times striking them from different sides is subject to statistical variations. Just as you do not always get exactly 10 heads and 10 tails when you toss a coin 20 times, in any given millisecond the pollen grain might be struck by slightly more atoms on one side than the other. For objects large enough to see with the naked eye, however, Brownian motion becomes negligible. Heisenberg uncertainty is a bit like Brownian motion at a yet smaller scale, as if atoms themselves were being knocked around by particles even tinier and harder to discern.

So, where are the famous conceptual difficulties of the quantum world? All the phenomena we have encountered so far—the two-slit experiment, Heisenberg uncertainty, even the dreaded quantum collapse—can be explained merely by postulating some kind of fine structure to space that is too delicate to measure directly, at least with present-day instruments. This hypothetical fine structure (the technical term for it is "hidden local variables") supports waves that can influence the motion of both photons and more solid particles and make small objects judder about so as to complicate the measurement of their positions and motions. Abrupt collisions jolt particles loose from the waves they are currently associated with.

We are doing very well at drawing a purely classical picture of quantum behavior. Where has the weirdness gone?

CHAPTER 3

COLLAPSE BY INFERENCE

If observing or measuring a particle involves doing something physical to it, then it is believable that such observation always has an effect on the particle, "knocking it off its guide wave" in the picture we have been trying to construct. So far, however, we have considered just two kinds of measurement; photons or other particles hitting a wall of detectors at the back of a two-slit experiment, and in the case of particles heavier than photons—electrons or oven-ready chickens—spraying light on them from an external source while they are still in flight through the experiment. Obviously, many other kinds of measurement are possible.

One option in the two-slit experiment is to respect the privacy of the particles while they are in flight, but place detectors at each of the slits to record which slit they pass through. If the particles are large things like bowling balls or oven-ready chickens, you can imagine all sorts of simple gadgets that could do the job—a lever that the object pushes as it passes, a beam of infrared light that it interrupts, a weight-detecting platform, and so on. If the objects are small things like electrons, the technology becomes a bit more subtle, but there is still a range of choices: various different electrical and magnetic effects can be used.

By now you will probably not be surprised to hear that in fact, placing such detectors at the slits destroys the interference pattern. When you think about it, any kind of detector cannot avoid doing something to a particle passing it—hitting a lever slows it down, shining a beam of light on it gives it a slight push, and so forth. Presumably the particles are getting knocked off their guide waves by their interaction with the detectors.

But now for the twist. What if we place a detector by just *one* of the slits—say, the left-hand one? Electrons going through the left-hand slit will no doubt be knocked off their guide waves. But you might reasonably suppose that if an electron goes through the right-hand slit, it will carry right on surfing. In that case the results at the back wall of detectors should be intermediate between those of Figure 2-1 and those of Figure 2-2. Half the electrons should arrive still riding waves, and therefore contribute to a partial interference pattern.

But what happens is that the results are exactly as shown in Figure 2-1. The mere presence of the detector at one slit completely abolishes the interference pattern—even though the detector does absolutely nothing, and registers nothing, in the case of electrons that pass through the right-hand slit. It would appear that the statement, "Measuring which slit the particle goes through knocks it off its guide wave" is to be taken literally—*even when the knowledge gained is of an inferential kind*, because of course we do not need two detectors to know which slit every electron passed through. If our electron detector clicked, it was the left-hand one; if it did not click, then by logical deduction, it was the right-hand one.

This is disconcerting, but there is still a way to cling to the classical picture. Any kind of detector—even of the most passive sort—has some effect on its surroundings, even when it is not detecting anything.[1] Just possibly, even the most innocuous detector somehow disrupts any guide waves passing nearby, which explains why a detector beside one of the slits is sufficient to destroy the whole of the interference pattern.

It gets worse, though. So far, we have considered only the behavior of isolated particles. In terms of our surfer analogy, each surfer has been doing his own thing, riding his own guide wave, and ignoring everybody else. This is a good approximation for photons, which are lightweight compared to solid matter and do not normally interact significantly with one another. We can think of each photon as riding its own guide wave, and the guide wave being sculpted by the bulk matter—walls, mirrors, and so on—with which it comes in contact. It is also a good approximation for isolated electrons that are flying through a vacuum. But these are rather special cases. It's time to consider what happens when particles interact.

We'll start with a simple example. Suppose that two electrons are fired from opposite sides of a vacuum chamber. If the trajectory of each is not known with perfect precision, that uncertainty will be greatly increased after they undergo a near collision in the center of the chamber. As they approach the center point, they will repel one another strongly, and as any pool player knows, the tiniest difference in alignment can make the difference between the particles rebounding toward their starting points, or being deflected sideways at some large or small angle. After the collision, both electrons will be flying out from the center in opposite directions, but there is no telling in which directions. We can regard them both as riding a circular guide wave that expands outward from the center of the chamber like a ripple. The guide wave behaves in the fashion we have come to expect —for example, it will generate an interference pattern if we make it pass through a pair of slits.

But when one of the electrons eventually gets measured—for example, by hitting a detector we have placed somewhere in the chamber —something very remarkable happens. Because the electrons are traveling in opposite directions, measuring where one of them is also tells us where the other is. Measuring one of the electrons also knocks the other one off its guide wave!

The technical term for such a relationship between two particles is *entanglement*, and it crops up rather often. Indeed, not just two particles, but a whole slew of them, can quickly become entangled. Imagine a boxful of electrons or atoms bouncing about like balls on a pool table. They are all riding their guide waves, and the possible arrangements tend to get ever more convoluted. The guide waves seem in some sense to be trying out every possible game of atomic pool that could theoretically take place. But examining just a few of the atoms—flashing a light on one corner of the pool table, so to speak—knocks all of them off their guide waves, effectively causing all of them to revert to behaving like particles. Is this kind of indirect effect capable of an ordinary physical explanation?

Actually, it has quite a good classical analog. Imagine a blind scientist investigating the properties of waves using a ripple tank, a shallow tank of water that is agitated to create patterns of waves on the surface. (These devices still exist, and were the best way to study wave patterns until modern computer simulations overtook them.) Because he cannot see the surface, the scientist has scattered smooth plastic beads that float on it and move with the ripples. He feels for the beads' position with his sensitive fingertips and thus performs useful measurements.

Unknown to him, however, the thermal control system, which is meant to keep the water at exactly constant temperature, is malfunctioning and causing the water temperature to drop below 0°C. Now it is a surprising but well-known fact that water that is very pure, containing no grains of dust or similar impurities to act as seeds round which ice crystals might start to form, can remain liquid at well below its usual freezing point. As soon as any such item is inserted, however, the entire volume of supercooled water turns almost instantaneously to ice.

This kind of instant freezing (physicists call it a phase change) appeals to me strongly as a metaphor for quantum collapse. For example, if we use a fluid that forms crystals with a well-defined orientation, the question "In which direction does the axis of orientation of the crystals lie?" has no meaning while the substance remains liquid, just as a quantum system has no specific state before measurement. It is making the measurement—touching the surface of the liquid with the tip of some instrument—that brings the definite orientation into being.

The relevant point here, however, is that the freezing is contagious, the state change rapidly spreads out through all the liquid in the vessel. This point inspired Kurt Vonnegut's famous satirical novel *Cat's Cradle*, in which a deranged scientist flings a seed crystal of the imaginary Ice-9 into the sea. The crystal triggers all the world's seas to turn solid within seconds, the effect rapidly propagating up even to semi-isolated bodies of water like the Great Lakes.

Back to the blind scientist: Each time he reaches out to feel for the position of the beads in the ripple tank, as soon as his fingertips touch the surface of the water the whole tank freezes instantly. So from his point of view, he always discovers the beads stationary in specific positions—yet those positions always form a mathematical pattern consistent with their having been propelled about by ripples until that moment. This is really quite similar to what happens when a quantum system is examined.

☜☜☜

So far, so plausible. But unfortunately the contagious collapse effect can be made even more startling. The key is very simple. Only very small systems normally show quantum wave-guided behavior we can easily detect, but what if we take the component parts of such a system, and separate them by a real-world-noticeable distance—an inch, perhaps, or even a mile? The mathematics of quantum predict that when we measure one of a pair of entangled atoms, thus allowing us to infer something about the other atom's position and state, the inferential knowledge gained knocks the other atom off its guide wave—and this happens *instantly*, however far apart the two atoms are at that moment.

At first encounter this sounds not just improbable, but impossible. In a relativistic universe, a signal that travels faster than light can also travel backward in time. But if this instant collapse did not happen, a terrible hole would open up in the fabric of quantum mechanics. By making simultaneous measurements on each of a widely separated particle pair, you could gain more knowledge about them than Heisenberg's uncertainty principle allows. This implication of faster-than-light "spooky links" between entangled particles is called the EPR paradox, after Einstein, Podolsky, and Rosen, who predicted the effect a lifetime ago. It led Einstein to believe that quantum theory must be wrong, or at any rate incomplete. It would seem that some kind of influence between the two atoms must travel faster than light, and in a relativistic universe a signal that travels faster than light can be used to send a message backward in time. In terms of our surfer analogy, it might seem that we could send the message just by kicking one surfer; his telepathically linked twin would instantly clutch his head and fall off his surfboard. In practice, sending a faster-than-light signal is not so easy as that. The basic problem is that the intended recipient of the signal has no way to inspect the second surfer without instantly making him tumble off his board anyway.

Indeed, turning Einstein's original thought experiment into a doable laboratory test turned out to be immensely hard. Quantum limitations apart, tiny things like individual atoms and photons are tricky things to measure in the hot, noisy environment of the Earth's surface. The breakthrough came when physicist David Bohm, whom we will meet again in a later chapter, described how "spooky links" could best be demonstrated, not by trying to measure the positions of two particles simultaneously, but in a more subtle way. In the 1960s, John Bell, a remarkable physicist whose day job was designing equipment for CERN, Europe's institute for particle physics research, developed Bohm's original proposal into a foolproof test.

While all particles possess the attributes of position and velocity, most also carry some internal information. Electrons have a property called spin, and photons a property called polarization. Spin and polarization are not the same thing, but they have a lot in common. Each can be regarded as a little arrow attached to the particle pointing in some arbitrary direction. In our surfer analogy, the surfer could indicate his polarization or spin by holding his arms at a particular angle as he stands on his board. Just as Heisenberg's uncertainty principle says that we can never measure both the position and velocity of a particle precisely, we are forbidden from ever measuring the exact direction of the arrow. We can get only a yes or no answer to a question about polarization or spin. It is as if the surfer can fall off his surfboard to the left or to the right, but give us no more than this hint as to what angle he was originally poised at.

What does this mean in a practical experiment? It is certainly possible to *produce* a photon that has been polarized at a particular, precise angle. In fact it is easy, for this happens whenever light passes through certain types of transparent material—for example, the lens of a pair of Polaroid sunglasses. If you take off your sunglasses and hold them at an angle of, say, 22 degrees to the horizontal, you can think of all the photons of light that pass through the lenses emerging with their little attached arrows pointing at just this angle.

The polarization of a photon is also easy to measure—it is far simpler than measuring any other property of fundamental particles. Again, the only equipment you need is another polarizing filter. You might think of the filter as a sort of portcullis gate that makes it likely that a photon whose arrow is pointing roughly parallel to the bars of the portcullis will slip through unscathed. What happens to those photons that don't get through depends on the type of material chosen for the filter. Sunglass lenses absorb those photons that don't make it, but in the laboratory we more usually choose a material that reflects the photons that are not transmitted, so that we can measure both sets if we wish. But the key point is that one photon can give us only one bit of information about its polarization. Like a yes-or-no answer, it can either get transmitted or not.

It is meaningful to speak of an individual photon being polarized at an exact angle, say 22 degrees, because this means that this is the only angle at which we can set a second polarizing filter that makes the photon certain to pass through it. The photon is also certain *not* to be transmitted if it hits a filter rotated 90 degrees from this direction. At intermediate angles the probability that the photon will pass through unscathed is given by $\cos^2\theta$, where θ is the angle between the photon polarization and the direction of the filter's axis. Whatever happens to the photon, the interaction with the filter resets the angle of its arrow, so that single bit of information—transmitted or reflected, which we can record as the number 0 or 1—is all the information about its polarization that we can ever actually read from an individual photon. All else is supposition.

Now we know everything necessary to understand perhaps the strangest experiment in the history of science, first performed in a foolproof way by Alain Aspect and colleagues in the 1980s.

The Bell-Aspect Experiment

Although photons do not normally interact much with one another, certain reactions can eject a pair of photons that travel in opposite directions but are entangled in the sense that their angles of polarization match exactly—even though an observer can never know precisely what that angle is. We know that if one of the photons hits a polarizing filter, it will be either transmitted or reflected. If it is transmitted, its angle of polarization changes to the same angle as that of the filter; if it is reflected, its new angle of polarization is exactly at right angles to that of the filter.

The rules of quantum guide waves tell us that at the moment one photon hits a polarizing filter, the polarization of the other photon instantly copies the change—to match the angle of the filter that its twin has just met. If we set the filter that the left photon is about to hit to 22 degrees, we immediately force the polarization of *both* photons to change to either exactly 22 or exactly 112 degrees, depending on whether the left photon is transmitted or reflected. In terms of our surfer analogy, this is much more subtle than knocking one surfer off his board and making the other follow suit. We are instead forcing one surfer to lean at exactly one of two possible angles, knowing that this will make his twin instantly twist to exactly the same angle.

It does seem like we have invented a faster-than-light communicator! To set up this useful device, imagine a spaceship orbiting slightly earthward of the midpoint between Earth and Mars, which we will assume are currently 200 million miles apart so that light takes approximately 20 minutes to travel between them. The spaceship has an apparatus for emitting polarization-correlated photon pairs, which reach Earth and Mars respectively about 10 minutes later, but with the Earth one arriving just before its Martian twin.

To send you an instant message, I will try to signal a 0 by holding up a polarizing filter in either a vertical or a horizontal position—it makes no difference which; both photons will be forced to a polarization angle of either exactly 0 or exactly 90 degrees, but I have no control over which of those values will be adopted. To signal a 1, I will hold up the filter at an angle of either 45 or 135 degrees. Again, it makes no difference which; both photons will be forced to either 45 or

TABLE 3-1 First Attempt at Faster-Than-Light Signaling

Filter Relative Angle	My Photon Is	Your Photon Is	Overall Result at This Angle
0	Transmitted	Transmitted always	
	Reflected	Reflected always	
			50/50
45	Transmitted	Transmitted 50%, Reflected 50%	
	Reflected	Transmitted 50%, Reflected 50%	
			50/50

135 degrees polarization. A moment later, you receive the other photon on Mars, and measure its polarization. If it turns out to be either vertical or horizontal, you write a 0; if it is slanted at 45 or 135 degrees, you write a 1.

Unfortunately, there is a snag in the scheme. You have no way to measure the polarization of the second photon exactly; you can only observe whether it is transmitted or reflected by your filter. Have a look at Table 3-1.

Whether I am holding my filter at 0 or 45 (or 90 or 135) degrees, the chance that your photon will be transmitted as opposed to reflected remains exactly 50 percent. The system is completely useless for sending messages. Of course there is a certain correlation between events at either end—if I hold my filter at the same angle as yours, the two photons always behave in the same way; if my filter is at 45 degrees to yours, they might behave differently, but this correlation only becomes apparent afterward, when we meet up to compare results.

There doesn't have to be any kind of link or conspiracy between the photons to produce the correlation. If each photon independently follows the rule "If I meet a filter at the same angle as my polarization vector, I get transmitted with 100 percent probability; if I meet a filter at 45 degrees to my vector, I get transmitted with 50 percent probability; if I meet a filter at 90 degrees to my vector, I get transmitted with 0 percent probability," then we get exactly the results in the table. To quote a well-known metaphor, it is no more surprising than opening a suitcase containing a right-hand glove and instantly deducing that your partner's suitcase must have the left-hand one.

That is really rather disappointing. A machine for sending signals faster than light would be most valuable, as would one for sending signals backward in time; it would be nice to be able to place a really sure bet on tomorrow's horse race. It must be worth another try. Let us try some more angles, shown in Table 3-2, and see if we spot anything promising.

By now, it should not surprise you that the final column is stubbornly 50/50 every time. Whatever I do with my filter, it is pure chance whether your photon is transmitted or reflected. No faster-than-light sending of information is permitted; we must forget that sure bet on the horses.

The interesting bit, however, is the figures in the other columns. At first sight they look quite innocuous. But hang on one moment . . . how does the universe know the figure in column 1, the relative angle between the two filters? My photon discovers the orientation of my filter when it bumps into it, and your photon discovers the orientation of yours, but neither should, on a classical picture, know anything about the orientation of the other filter, which is necessary to know the *relative* angle.

John Bell realized that if the photons are acting independently and always act oppositely when the filters are at 90 degrees, then the probability of getting an opposite result at smaller angles of difference should always be at least in proportion to that angle. His proof is now called Bell's inequality, and involves slightly arcane mathematics, but it is also capable of visual portrayal. Indeed, we met it in Chapter 1, and Figure 3-1 makes the link.

If the two lottery cards of the conjuring trick are replaced by polarizing filters that are dialed to the positions of the spots my partner and I pick, then however the photons are internally programmed—by whatever rules the lottery card color gets filled in—it should be impossible to choose angles 6 degrees apart and yet get the same result 99 percent of the time, as happens in the conjuring show and also with

TABLE 3-2 Second Attempt at Faster-Than-Light Signaling

Our Filters Are at Relative Angle	My Photon Is	Your Photon Is Transmitted with Probability %		Overall Result at This Angle
0	Transmitted	100		50/50
	Reflected		0	
6	Transmitted	99		50/50
	Reflected		1	
12	Transmitted	96		50/50
	Reflected		4	
18	Transmitted	90		50/50
	Reflected		10	
24	Transmitted	83		50/50
	Reflected		17	
30	Transmitted	75		50/50
	Reflected		25	
36	Transmitted	65		50/50
	Reflected		35	
42	Transmitted	55		50/50
	Reflected		45	
45	Transmitted	50		50/50
	Reflected		50	
48	Transmitted	45		50/50
	Reflected		55	
54	Transmitted	35		50/50
	Reflected		65	
60	Transmitted	25		50/50
	Reflected		75	
66	Transmitted	17		50/50
	Reflected		83	
72	Transmitted	10		50/50
	Reflected		90	
78	Transmitted	4		50/50
	Reflected		96	
84	Transmitted	1		50/50
	Reflected		99	
90	Transmitted	0		50/50
	Reflected		100	

Probabilities are rounded to the nearest percent.

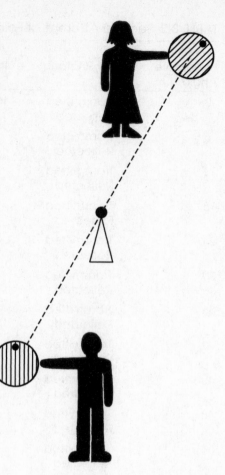

FIGURE 3-1 The lottery cards
shown as polarizing filters.

real-life photon pairs. There really does seem to be some spooky link
between the photons, across whatever distance of space, causing correlations that are otherwise inexplicable. No kind of hidden local vari-
able theory can explain this behavior. If we are still clinging to the
particles-plus-guide-waves story, we must also assume some kind of
faster-than-light link between the particles.

There are other ways to look at things, of course. We have encountered two mysteries in this chapter. The first was that observing one
slit of a two-slit apparatus can seemingly change the behavior of particles that go through the other slit, an arbitrary distance away. The
second was that observing one photon of a correlated pair can seem-

ingly change the behavior of its partner, an arbitrary distance away. Occam's razor suggests that we should seek a single explanation for both mysteries. One such hypothesis is this: "The acquisition of knowledge about a system by an observer, *even inferential knowledge,* can somehow change the behavior of that system—or at any rate what the observer subsequently sees—in a way unprecedented in classical physics, where the observer plays no special role."

CHAPTER 4

A HORROR STORY WRIT LARGE

The experimental results described in the preceding chapter are certainly surprising. But you may well be wondering why they have caused such a huge upset in science, to the point where some of the human race's most intelligent minds have been prepared to seriously consider wild philosophical ideas like those described by Professor Cope in Chapter 1. So-called observer effects are disconcerting, but they normally affect only tiny things. Quantum effects of every kind normally average out to produce large-scale behavior that obeys classical statistics; your Polaroid sunglasses, for example, reliably shield your eyes from the glare by absorbing an exactly predictable fraction of the photons that reach them, spooky quantum behavior notwithstanding.

If quantum does not actually cause a paradox at the macroscopic scale—as would perhaps be the case if you could transmit real information faster than light and backward in time—the layperson could be forgiven for asking: Can we not just overlook the oddities? Less excusably, many physicists take somewhat the same line, overtly or tacitly. This chapter is devoted to showing how the effects of quantum can be magnified, naturally and artificially, to the point where no one could possibly ignore them. As we go, we shall make a list of what I will call the PPQs—the Principal Puzzles of Quantum.

It Can Be a Big Deal

First, let us demolish the idea that quantum weirdness only ever affects microscopic systems. Quantum effects can be amplified quite easily. The lottery cards of Chapter 1, for instance, are not just a metaphor; they could be manufactured for real. Each card would have to contain some mechanism that created a quantum-linked system when the card was torn. In principle, that could involve a pair of photons exactly like those in the Bell-Aspect experiment; each photon could be stored in an arrangement called a high-finesse cavity, shuttling to and fro between two almost-perfect mirrors. That is technically difficult; a more promising approach would store the link in the spin of quantum-correlated atomic nuclei. When either half of a card was scratched, a mechanism could just measure the magnetization of the local nuclei using the technique called nuclear magnetic resonance (NMR for short) and release a chemical that would turn the relevant spot black or white depending on the result. Using that technology, the cards could be made reasonably small—my guess would be about the size and mass of a pocket PC. They would be expensive, but they would work just as described in the story.

Once a measurement is made, its consequences can always be amplified indefinitely. One possible objection to the test described in Chapter 1 in which one-half of a lottery card is sent to Australia, and then both halves are scratched and measured simultaneously by machines so that there is no time for any speed-of-light message to pass between the cards, might be on these lines: Perhaps the color of the lottery card does not really turn properly black or white until a fraction of a second after the measurement is done. For example, if scratching the card triggers a chemical reaction, it always takes a little time for a stable compound to form. An analogy is one of those fairground games where you must throw a ball onto a tray of bottles with funnel-shaped necks. The ball bounces around tantalizingly between one bottle and another. Sometimes, even after it appears to have made its choice and is rattling around in the neck of one particular bottle, it can still spring across to a neighboring one at the last minute. Perhaps the color of the spot is not truly determined until there has been time enough for speed-of-light signals to bounce to and fro between America and Australia.

To answer this argument, we will scale up the lottery-card experiment to a version where you are on Earth and your partner is on Mars. We will scratch the lottery cards when the planets are far apart in their orbits, so that light takes about 15 minutes to get from one to the other. Moreover, we will assume that your partner has a morbid fear of the color black, so if her spot turns out to be black, she will immediately shoot herself.

You each scratch your card. On Earth, your spot turns out to be white, and you know that if the cards work, your partner is 99 percent likely to be safe. If we are in a classical universe, with no faster-than-light signaling of any kind allowed, your partner's spot has no way to know this and there is presumably a 50 percent chance that it initially appear black. It cannot know about the result of your measurement for 15 minutes. Do you really believe that all the molecules in the gun, the bullet, your partner's body, and so on, had not quite decided which positions to be in for a quarter of an hour? And of course you could have scaled up the outcome on Mars (or indeed, Earth) even further, with a device that would trigger an H-bomb if the card turned out to be black, and so on. There is no limit to the amplification that you can do.

A Virtual Time Machine

I have claimed that, in a relativistic universe, being able to send information faster than light implies that you could also send it backward in time. This is not the place to explain special relativity fully, but I want you to feel this point in your bones, and we can describe the essentials quite simply.

Einstein realized that if the speed of light is the same for all observers (the basic assumption from which all of special relativity can be deduced), then the sequence of events can appear different to different observers. We will explore this with a slight extension of his original thought experiment using a railway train as shown in Figure 4-1.

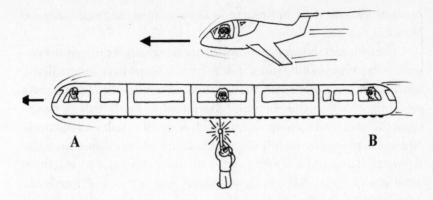

FIGURE 4-1 At the midpoint of the train are three observers—one on the train, one by the trackside, and one on an airplane that is overtaking the train. The observer on the train thinks A and B receive the light signal simultaneously. But the trackside observer thinks that B gets the signal first, and the aviator thinks that A gets it first.

Suppose a lamp mounted at the center of a train flashes. Clearly, two observers on the train, one stationed, say, exactly two cars ahead of the lamp, the other exactly two cars behind it, will see the flash at precisely the same instant. It makes no difference whether the train is stopped or moving.

But we have an apparent paradox if the train is indeed speeding along, and we consider the point of view of an observer who is stationary with respect to the Earth. For convenience, let us suppose that he is standing beside the track at the point where the light is flashed. Because the front of the train is receding from the light pulse, whereas the rear is advancing to meet it, he will unambiguously measure the flash as reaching the rear observer on the train before it reaches the front observer. The difference would be tiny—on the order of 10^{-13} seconds for a real train—but it can be much larger if we are talking about faster and more widely separated systems, such as imaginary spaceships or real stars or planets moving at high relative speeds. Conversely, from the point of view of an observer moving the other way with respect to the train—say, a pilot overtaking it in an aircraft—the rear observer on the train receives the pulse after the front observer. How can this be?

The different sequences witnessed are all equally "real"—some observers can quite validly think that A got the signal first, others that it was B. Luckily or otherwise, though, no one can use the fact to send a signal backward in time because, considered as two events in space-time, the time-and-space point at which A sees the flash and the time-and-space point at which B sees the flash are what is called spacelike separated. This means simply that the spatial separation between them is sufficiently great that it is impossible for any light-speed message to pass from one to the other, in either direction, in the time interval between the two events. This applies from the point of view of any observer. For example, an alien in a fast spacecraft overtaking the train at 99.99 percent of the speed of light will see the train contracted to a tiny fraction of the length it appears to us, and will measure the front observer getting the flash significantly before the rear one, but still without enough time passing for anyone to take a message from the front observer to the rear one in the delay between the two events. No signal can pass between two spacelike separated events in the time available, so neither event can possibly cause the other, or indeed have any effect on the other.

It turns out that any two events in space-time are always unambiguously either spacelike separated or timelike separated from the point of view of all possible observers. If they are timelike separated, then one can have influenced the other, but the order is always unambiguous. In our normal world, the difference is usually very obvious. For example, the events of Columbus setting foot in America and your picking up this book are timelike separated, and Columbus unambiguously happened first: Columbus's actions might have had an effect on you, but not vice versa, and any alien observers zooming spaceships around in complicated patterns will agree with you on this point.

So the problem that widely separated events may appear to happen in a different order to observers moving at different speeds is purely one of bookkeeping. Back in the Victorian era, when the first transatlantic telegraph wires were laid, people found it very puzzling that they could send a message from London to New York that could be physically delivered in New York (normally by a telegraph boy on a bicycle, clutching a typewritten sheet) at a time before it had left London, as measured by local clocks in each case. When telegraph wires were laid round the world, even across the Pacific, might it have been possible to send a message around the world to yourself that would arrive before you had sent it? Of course, intelligent people realized that this was nonsense, but before the position of the International Date Line was agreed, the point caused considerable confusion. Jules Verne had fun with these difficulties in *Around the World in Eighty Days*, and Oxford mathematician Charles Dodgson (best known for his books, *Through the Looking Glass* and *Alice's Adventures in Wonderland* under the pseudonym Lewis Carroll) amused himself by sending spoof enquiries to the telegraph companies about the matter. Nowadays, we all know that claims like "If you fly from London to New York by Concorde you will land before you take off" merely refer to clocks set to different time zones. Of course, you are not really traveling backward in time. Similarly, observers on differently moving spaceships inferring by subsequent observation that distant events happened in different sequences is in no sense a real paradox.

But if you *could* somehow send messages faster than light, this sequencing problem suddenly would become real. To see how, look at the position in Figure 4-2, where two very long and fast trains are passing in opposite directions. Both trains are equipped with instant

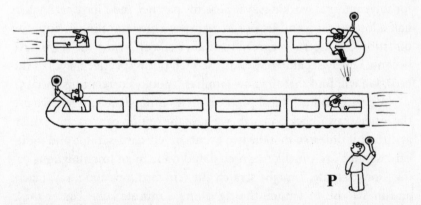

FIGURE 4-2 If the conductor on each train has a device that allows him to send a signal to his engineer instantaneously, P gets his message back before he sends it.

signalling devices linking the engineer at the front with the conductor in the caboose at the rear. You are standing at position P. You ask the conductor of the train passing you if he would be so kind as to send a message to his engineer for you.

The conductor obligingly sends the message, which travels instantly in his frame of reference. As we have just seen, in your trackside frame of reference, it gets there a little earlier than it set out—for convenience, let us say 1 second earlier, though that would be more realistic with spacecraft than trains. There is no obvious paradox yet, but now suppose the engineer of the leftward-going train sends your message over to the conductor of the rightward-going one (he can do this by ordinary slower-than-light signaling, because the trains are close together), and asks that conductor to send the message on to *his* engineer using his own faster-than-light signaler. Once again, in your trackside frame of reference, this signal arrives a second before it was sent. You get your message back 2 seconds before it was transmitted! And now all the familiar paradoxes of time travel arise. For example, what if the message asks the engineer of the rightward-going train to shoot you—therefore preventing your sending the message that asked him to do this?

We do not really have such an instantaneous communicator, but what we do have is an unlimited supply of lottery cards that (unless we adopt the extreme philosophical positions described by Professor Cope) seem to require an instantaneous communication mechanism for their internal workings. Suppose my partner and I have each taken half a lottery card, and set out in two spaceships traveling in opposite directions. At a certain agreed time after takeoff, when the ships have become widely separated, we each scratch our respective halves of the card. We will find that the now familiar "spooky correlations" occur.

A pretty philosophical problem arises immediately. When we use the lottery cards back on Earth, we have the option of scratching them at different times, so that the two events of left-card-scratch and right-card-scratch are timelike separated and done in an unambiguous order. For example, I might scratch the left card, revealing, say, black, and invite you to scratch the right one a minute later. Under these circumstances, presumably my card acted as the master, decided which color to be itself, and sent some signal to your card telling it how to behave subsequently. But in our spacelike-separated spaceships, because we are traveling in different directions, from the point of view of a leftward-traveling observer, I scratched off my card first—it was the master and yours the slave—whereas from the point of view of a rightward-traveling observer, your card was scratched first, and mine had to conform with it. Because in special relativity, no frame of motion is better or more correct than any other, there is no way to answer the question of which card influenced which. But for many classical physicists, a far more troubling puzzle is this: From some points of view, whichever card acted as master sent a signal that *retrospectively* determined the outcome at the other card's location. How can this possibly be?

To make the horror of the lottery cards clear, a classic science fiction story that I read as a teenager illustrates the point rather vividly. In the story, a conventionally minded physicist is sent to investigate an alleged case of psychic powers. The subject (who appears completely unaware of her own spooky abilities) is a hospital patient in an isolation ward, a blind lady whose only news from the outside world comes from a nurse who reads her randomly selected stories from the local newspaper. The nurse has noticed a strange thing: Whenever she reads the old lady a sob story, it turns out subsequently to have a happy ending, even in circumstances where that seemed very improbable. If the blind lady is read a story about an abandoned baby, the mother later has a change of heart and returns lovingly to collect it; if it is about a cancer sufferer, the person goes on to have a spontaneous remission, and so forth. None of the instances taken on its own is in any way impossible, just lucky, but the odds against this happening for every story the blind lady is read are overwhelming.

At first, the physicist is extremely skeptical. But after many increasingly foolproof tests, he is driven to the conclusion that the old lady does have some kind of psychic power: She can heal other people and situations. Very reluctantly, he accepts that she must be able to perform some kind of unconscious action at a distance, and he is able to integrate this fact into his worldview.

Then the old lady hears a story about an air disaster that happened a week earlier. The crash happened in remote mountains, and the site has not been found, but there is no realistic hope of survivors. The day after she hears the story, the entire crew and passengers from the airliner limp into a remote village, weary but not seriously hurt. They report that by a million-to-one chance the plane bounced from trees into a snowdrift, without injuring anybody, but in such an inaccessible location that it has taken them this long to make their way to civilization.

The physicist's hair stands on end as he tries to work out how this can be. Did decaying flesh and bones slide about, reassembling themselves into intact, healthy human beings? Or—in a way even more terrifying—could the old lady's power reach backward in time, undoing events that had already happened? The scientist spends the rest of his life trying as hard as he can *not* to think about what really happened on that mountainside. This is really quite reminiscent of some physicists' attitude to quantum paradoxes.

Tiny Particles Make Huge Waves

Another point I want to emphasize about quantum is the sheer gigantic size the wave associated with every particle can grow to. The two-slit experiment is normally performed in a container roughly the shape and size of a shoebox. But of course the wave associated with a single photon can explore not just two, but an infinity of routes, and over unlimited distances. For a more dramatic illustration, consider the Temple of the Photon, a place I have just invented—although I have been in a Manhattan restaurant whose decor resembled it alarmingly. The Temple of the Photon is a cathedral-like open space with a great complexity of randomly placed pillars, statues, bas-relief sculptures, and so forth. Its distinctive feature is that every surface is coated with a perfectly reflective substance. The only exception is a square canvas on one wall, which is coated with ultrasensitive photographic film. Into the temple we take a low-intensity photon source, which we leave for a week or two. Photons will be emitted at an average rate of one per second, and whatever trajectory each one follows, it will eventually strike the photographic film, because that is the only place it can be absorbed.

When we develop the photographic film, we will see a very complex interference pattern, far more convoluted than the simple stripes of the two-slit experiment. (It will more closely resemble a hologram, which is made in quite a similar way.) But the pattern will include lighter areas and darker areas, and typically some spots that are perfectly black. The only way to work out the pattern, and in particular, where those black spots of perfect cancellation occur, is to trace every possible path that the photon could take from its source to that point on the film, and calculate the length of each trajectory to an accuracy much better than the wavelength of light. The sum of the interference effects from all the infinity of slightly different paths tells us whether the spot will be dark. Changing anything in the building—moving a small statue in one of the side aisles a fraction of a micrometer, say—changes the position of the dark spots. As the pattern builds up, one photon at a time, each and every photon must explore the whole temple—trace every possible trajectory through it—to decide where on the film to alight. If even a few photons missed out on exploring even some of the possible trajectories, they would not know to avoid the dark spots, and the pattern would be contaminated.

It gets more extreme than this. Imagine that back in the early universe, an atom emits a photon. The photon travels through space for 13 billion years, until it eventually strikes the mirror of the Hubble space telescope and expires against an electronic detector, contributing to one of Hubble's long-exposure, deep-sky pictures. For 13 billion years, that photon has been riding an expanding wave bubble that has mapped out a volume of 10^{31} cubic light years—all to correctly guide the trajectory of one tiny photon. If the atom emitted a couple of photons in rapid succession, their polarization might be linked, just as in the Aspect experiment. If the Hubble telescope happens to be using an instrument with a polarizing filter, then its measurement effectively causes another photon 26 billion light-years away—far beyond the currently observable universe—not only to "fall off its guide wave" into a specific location, but to do so in a way that correlates with Hubble's measurement. You might say it forces that incredibly distant surfer to tumble off his board at a particular angle, which we can control to be one of two choices.

The resultant effect does not require cleverly designed lottery cards to amplify its results to macroscopic significance. Systems that involve repeated collisions—whether of air molecules or real-sized billiard balls on a baize-topped table—multiply very small initial effects in an exponential way. This is why many big-money lotteries use a tumbling cylinder of balls to determine the winning numbers: The position of individual balls rapidly becomes completely unpredictable.

Long before chaos theory was invented—in fact, back in 1914—a physicist named Borel demonstrated mathematically that the minuscule change in gravitational attraction caused by moving a small stone a hundred light-years from Earth a few centimeters would completely change the positions of all the individual air molecules within our atmosphere a few seconds after the field alteration reached us. The famous butterfly effect then takes over; tiny alterations in microscopic air currents totally alter the weather pattern of the whole Earth within a few weeks. And chaotic systems like the weather have a very significant effect on human history. If the Spanish Armada had not been scattered by a freak storm as it was on its way to attack England, subsequent European history would have been very different. A similar event affected the Far East a few centuries earlier, when a huge fleet sent from China to conquer Japan was also defeated by bad weather. Something as tiny as the motion of a single subatomic particle not only can, but usually does, alter the whole course of history.

So when a human-made telescope detects a photon that has been traveling through space for millions of years, or a cosmic-ray detector buried beneath Antarctica detects half of a smashed atomic nucleus that has been voyaging for a similar time, the result can have very real implications for events in, say, the Andromeda galaxy. Light from Andromeda takes a million and a half years to reach Earth. The Andromeda that our descendants observe a million and a half years from now will be seen to have evolved in, ultimately, an utterly different way—affecting perhaps such things as which planets in Andromeda do or do not develop life, and whether one spawns intelligent beings that go on to set up a galactic empire—according to which way we put the filter in our telescope.

In a final generalization of our original rather contrived (if potentially makeable) lottery cards, these long-range spooky influences affect not just things that have been in very intimate contact, such as photons or other particles that originally came from the same atom. These are just cases where the spooky link is easiest to observe in practice, as in the Aspect experiment. Any pair or larger group of particles that have once interacted—for example, two electrons that were once in the general vicinity of one another—will show a certain degree of spooky correlation in their subsequent behavior. Any measurement-style interaction with one has a subtle effect on the rest. The photons and other particles that enter Earth's atmosphere each second are thus directly and indirectly linked to just about every other particle in the observable universe. And when such a particle is measured by striking some terrestrial object, it seems to have some subtle instant effect on all other particles everywhere.

I stress once again that these links are not causative, in the sense that we cannot use them to send any kind of information or message. As with the lottery cards, we can measure—but we cannot force the result of a measurement. We cannot use these effects to explain alleged telepathy, for example. But the universe does in a certain sense appear to behave holistically, as if interactions in every part have subtle effects on every other, and if we did try to explain this behavior by some kind of built-in faster-than-light signaling mechanisms, then those mechanisms would by implication have to be capable of sending signals backward in time. Something strange is indeed happening. Here is our first Principle Puzzle of Quantum.

PPQ 1: Spooky quantum links seem to imply *either* faster-than-light signals *or* that local events do not promptly proceed in an unambiguous way at each end of the link.

This puzzle leads directly to another disconcerting feature, the intrinsic randomness of quantum. We have been talking about quantum outcomes, such as whether a photon is reflected or transmitted from a filter, as happening "randomly," but maybe you took that with a pinch of salt. After all, we call everyday events like spinning a roulette wheel or tossing a coin random, even though someone with a sufficiently clever little computer-and-radar kind of arrangement could predict the outcome. (Indeed, as I write, a scandal involving something similar at a real-life casino has just hit the headlines.) Could not the "random" part of a photon's decision which way to go when it hits a filter really just be some function of the way the molecules in the filter are bouncing about at the moment it strikes, for example?

It would seem not. For if we could ever force even a minor-seeming exception to true quantum randomness by tinkering with local conditions, a true paradox would follow. To see how, suppose we have found some lottery cards like those of Chapter 1, but just a tiny bit biased. When you scratch the left-hand card in a strong magnetic field, the probability that you will get white is 55 percent rather then 50 percent. You make a plan as follows:

"DARPA has offered us a fabulous sum if we can send a message faster than light with these cards," you tell your partner. "So we will take a stack of 1,000 cards and tear them down the middle: I will take the left half of the stack and you the right. I will scratch my cards in the presence of a magnet.

"DARPA will ask me to send you a single binary digit, which will obviously be either 0 or 1. If it is 0, I will scratch the top, 12 o'clock spot on each of my cards. If it is 1, I will instead scratch the 3 o'clock spot.

"You need only scratch the top, 12 o'clock, spot on all your cards. We know that if I am also scratching the 12 o'clock spot, your color will be the same as mine every time; on the other hand, if I am scratching the 3 o'clock spot, your color will be different every time. On average, 550 of my spots will be white in either case. So if I am scratching the same spot as you, you will see about 550 whites. If I am scratching a spot at 90 degrees to yours, you will see only about 450 whites. Tell the DARPA examiner the answer is 0 if you see more than 500 whites, 1 otherwise. The chance we will get it right is greater than 99.9 percent!"

A similar strategy could be devised if there were any quantum systems that in any way departed from the perfectly random behavior predicted by quantum mathematics. Using the loophole, you could indeed send a message faster than light, hence backward in time, with potentially paradoxical consequences. Quantum randomness appears to be truly fundamental, truly unpredictable. This is intuitively hard to accept, and inspired Einstein's famous comment that he could not believe that God plays dice with the universe.

PPQ 2: Spooky quantum links seem to imply *either* faster-than-light signals *or* that quantum events are truly random.

A third puzzle of quantum is the sheer baroque quantity of calculation the universe must apparently do to determine the outcome of each microevent. For example, the wave associated with the photon described above, emitted early in the history of the universe, seemingly had to explore every inch of billions of cubic light-years of space in order to decide where the photon would eventually alight. The task becomes still more impressive when we consider how clever such a wave sometimes has to be.

Remember the oven-ready chickens version of the two-slit experiment? If each chicken has a bar-code tag attached, then a detection of a chicken passing through a slit might be accomplished by placing a bar-code scanner and printer, as used in supermarkets, beside one of the slits. Each time a chicken flies through the slit, the scanner prints an appropriate line on the checkout roll, thus making a record of its passage in the form of a permanent impression on the surrounding environment. As we have discovered, placing such an arrangement by just one of the slits, say the left one, is sufficient to prevent any interference pattern from forming. The universe somehow knows to stop providing guide-wave interference for all chickens—even those that go through the right slit—once the detector is switched on.

In terms of our guide-wave hypothesis, it follows that the presence of the scanner must be disrupting the guide wave itself as it goes through the left slit. That is conceivable. Any detector has some effect on its environment—for example, a standard bar-code scanner would shine a tiny red laser beam across the slit, and it's plausible that the beam might disrupt the guide wave. But we can fine-tune the arrangement further. Suppose that we program the scanner to suppress printing when a 4-pound chicken passes through. Chickens of all other weights—3 pounds, 5 pounds, or whatever—are to be recorded as before, but there will be no way to tell that a 4-pounder has passed through by examining the checkout roll afterward. Now when we fire

4-pound chickens through the arrangement, we get a full interference pattern. But when we fire chickens of weights the scanner is programmed to detect, we get no interference. How can the universe possibly "know" to make all 4-pounders form an interference pattern, when all we have done is change the internal programming of a scanner that half the chickens (of every weight) do not even go near?

There seems to be only one logical answer. The guide wave must somehow be so clever that it tests the effect its associated chicken *would* have *if* it were to pass the bar-code scanner—putting the computer inside the scanner through its paces even though the chicken is passing through the other slit. The guide wave of any non-4-pound chicken thus discovers that it should disrupt itself when passing the slit.

Can the wave really be that clever? It seems highly implausible. But it is not impossible that the guide wave carries such detailed information about its associated particle in every part of it. An analogous object is nowadays familiar. A hologram contains its whole picture in each part of itself. You can test this by smashing a glass hologram and peeking through one of the fragments, or more safely and less expensively by covering up all but part of the hologram with a paper mask and examining at different angles the bit that remains exposed. Perhaps guide waves behave like that, temporarily fooling the universe in the same way that a hologram can deceive our eyes about the apparent position of an object, testing what would happen if the associated chicken's label were to pass the scanner. The guide-wave hypothesis survives, barely. Nevertheless, when we consider the potential hugeness of each guide wave in conjunction with its extraordinary cleverness, we are justified in formulating a third PPQ.

PPQ 3: Why does the universe seem to waste such a colossal amount of effort investigating might-have-beens, things that could have happened but didn't?

Another problem with the wave-rider picture that we have been trying to build is more subtle. So far we have spoken of wave-riding particles as undergoing two kinds of interaction. The first kind was an encounter with another wave-riding particle. The result of that is that each particle continues on its way, but riding a more complicated wave shape, and now with a curious relationship between the fates of the two particles, which we call entanglement. The second kind was a measurement, something that knocked the particle off its wave altogether, for example, hitting a solid wall of matter. But now we recognize that *everything* in the universe is just particles riding guide waves, the waves becoming more and more entangled as the particles repeatedly encounter others. So when does a definitive measurement ever get made?

My high school physics teacher had a rough-and-ready answer. Small particles typically have quite long wavelengths associated with them; an atom usually has a wavelength much longer than its own diameter. But large things usually have smaller wavelengths, much tinier than the object itself. Indeed, anything big enough to be seen with the naked eye has an associated wave that is ultramicroscopic. So measurement can be crudely defined as what happens when a little thing interacts with a much bigger one. The wavelength associated with a massive thing like a planet is almost unimaginably tiny, so a measurement interaction with an instrument on the surface of Earth gives a definite result "for all practical purposes," my teacher claimed.

His story sounded plausible. After all, there are many cases in which frenetic and complicated behavior at the small scale averages out to solid and predictable behavior at the large. Even classically, no one molecule in your body is sitting still. Each is bouncing around at a speed of several hundred meters per second. But when you are sitting still in a chair, the total average momentum of all those trillions of atoms divided by their collective mass is zero, or as near as makes no difference. If we think of measurement simply as what happens when a tiny thing encounters a much larger one, then it should be no surprise that the interaction makes for a more stable result.

To an extent, my teacher had a point. The position of Earth's center of mass is pretty well defined. However, Earth can potentially enter an enormous number of different states—for example, with different weather patterns on its surface—without affecting its position in space. Chaos theory tells us that there are many situations in which even the tiniest initial difference (whether a photon gets reflected or absorbed when it hits a water surface at an angle, for example) can multiply to produce a completely different worldwide weather pattern a few weeks later. There is no natural tendency for events always to converge in a single consensus pattern.

Once some one thing is decided for certain, there is a tendency for the rest of the world to fall into a specific pattern in a kind of domino effect, as when the blind scientist touched the surface of the supercooled water and triggered freezing, in our earlier metaphor. But given that everything in our universe, including scientific instruments and even our own brains, is composed of wave-riding particles, what can ever start the fixing process? The situation is a little like a children's party where Mary knows that she wants to sit next to Billy but avoid Susan; Joanna that she wants to be on Helen's right but far from Doug unless Jane is between them, and so on. People have an idea about the relative positions they want to occupy, but no one is prepared to be the first to sit down.

And so we come to our final puzzle. It appears that on the one hand the universe must be clever enough to keep calculating an enormous number of diverging possibilities for long periods (perhaps forever) and yet in some mysterious way produces a single actuality that we see as its output.

PPQ 4: Why does reality appear to be the world in a single specific pattern, when the guide waves should be weaving an ever more tangled multiplicity of patterns?

For convenient reference, you will find the four PPQs listed at the back of the book. But what status do these problems have? None is quite a paradox in the strict sense, and yet each somehow feels like it is more than just an aesthetic problem with the theory. The list is, in a sense, merely a personal one. It highlights the features of quantum that my physical intuition finds the most troubling. But I am in excellent company, because these problems also troubled the founding fathers of quantum, some of the greatest physicists who ever lived, including Einstein himself.

CHAPTER 5

THE OLD TESTAMENT

T his is not a history book—it is a book about new ideas and progress. But sometimes there are lessons to be learned from history and from failure. Dante justified writing the *Inferno*, far more readable than his corresponding books describing paradise and purgatory, with the claim that exploring evil is one way to learn the path to good. At junior school, a classmate of mine once asked our highly religious headmaster why the Bible includes the Old Testament, with its descriptions of so many wicked things. He replied after some hesitation that one reason was to show the contrast between Jesus's teachings and those of the harsher Old Testament prophets.

In something of the same spirit, we will now look at the traditional interpretations of quantum mechanics: those that originated in the first half of the 20th century, and remain (bizarrely) the best known to many science students today. I am not sneering at them, because it is easy to be wise with hindsight. But it must be said that they do not show the physics community in its best light. Please keep your skeptical instincts alert as you read on, because we are about to encounter stories of stubbornness, denial, and wishful thinking. Above all, remember that we should never believe something merely because it is advocated by someone who is very famous, or very well enshrined in history. If we took that attitude unquestioningly, we would still be endorsing the scientific beliefs of the philosophers of ancient Greece.

These cautions given, let us now look at the views of the founding fathers of quantum, some of the greatest scientists who have ever lived.

Schrödinger

Erwin Schrödinger developed the wave-theory formulation, which described the previously mysterious hydrogen atom with triumphant accuracy. Schrödinger's interpretation of his wave mechanics was as simple as it was bold. His answer to the problem of wave-particle duality was that there are no particles, only waves. Just as a tsunami wave may be spread out invisibly thinly in the deep ocean, but can rise and become concentrated as it passes over shallow water, ultimately depositing most of its energy on a narrow stretch of coast, so any kind of wave can vary greatly in its physical extent. Schrödinger thought that the apparent particles of radiation and matter were merely manifestations of waves squeezed to an extreme degree—as when a water wave focused by the shape of an estuary rears up to a sudden peak and expends all its energy in knocking down a tall lighthouse, for example.

Schrödinger's view works quite well for bound particles, such as electrons in an atom, whose behavior is described by the "time-independent" Schrödinger equation, which does not even try to answer the question of where the particle is located at any given instant. But it works much less well for particles in free space, such as an isolated proton or electron. Then the time-dependent version of the equation predicts that as long as it is not interacting with anything, the wave will continue to gradually flatten and spread out, in principle extending to infinity. Yet even a tiny observation-like interaction somewhere in this volume of space can bring an extremely pointlike electron springing into view, with dimensions that remain too small to measure—and this happens in less time than it would take light to cross the region of space where, until that moment, we thought the electron might be. As we have seen, it is hard to imagine any reasonable physical mechanism that could bring about quantum collapse in this kind of nonlocal case.

Schrödinger did not claim to have an answer to the problem, but he did make clear his contempt for the idea, which underpins several of the interpretations described below, that a system might be regarded as not having a definite state until an observation is made. If we accept that this notion makes sense in the context of microscopic systems, Schrödinger argued, then in the appropriate circumstances it would have to apply to larger systems as well—even living things such as cats. Suppose you constructed a completely observation-proof box and placed within it a cat and a sort of Russian roulette device which, as soon as the box was sealed, would fire a photon toward a polarizing filter and kill the cat if the photon happened to pass through. There is a fundamentally unpredictable 50 percent chance that the photon will pass through the filter. By the "nothing is actual until observed" argument, the cat would have no definite state until the box was opened, maybe hours later, to reveal either a dead cat with rigor mortis or a live but hungry one. Schrödinger invented his famous parable of the cat-in-a-box not to be believed, but to be disbelieved, as a *reductio ad absurdum*. He thought that it was manifestly ridiculous to think in the terms that the cat is neither dead nor alive until the box is opened.

Born

Max Born had an alternative way to look at Schrödinger's waves. He saw them as waves of probability. It will be useful to us later on to understand the modern philosophy of probability, and for this reason and for the sake of clarity, I shall extrapolate his argument into modern terminology and examples.

There is a subtle difference between probability and statistics. Consider the difference between the two following questions:

"There are 100 people in this hall. Fifty of them have had a white sticker placed on their backs. What percentage have white stickers on their backs?"

and

"I have selected at random someone from the hall who now stands before you. What is the chance that this person's back bears a white sticker?"

The answer to the first question is straightforward: 50 percent. The second question is trickier. After all, the person standing before you either has a sticker or doesn't. If everyone in the audience except you can see the subject's back, then everyone else in the room already knows that the hypothesis "There is a sticker on this person's back" is either true with 100 percent certainty or false with 100 percent certainty. In what sense can it be correct to answer "50 percent probable"?

The matter becomes even more puzzling when you consider that the probability can *change*. Suppose you hesitate to answer the second question and the host goes on to say, "I will give you some further information. There are 50 women in the hall and 40 of them had white stickers placed on their backs. The remaining 10 stickers were distributed among the 50 men."

You can see that the person before you is a woman, so it is reasonable to revise your estimate upward to 80 percent. But how can it be rational to do that? The person before you has not changed, nor has the fact that she either has a sticker on her back or has not. How can the right answer have changed?

The answer that most philosophers of mathematics would give is that probability is best thought of as a measure of ignorance. It is not rational for you to think that the physical facts of a situation change when you are given new information, but it is rational for you to take into account your reduction of ignorance. That this is not a trivial distinction is shown by the famous Monty Hall problem, in which a game show host shows you three cabinets, and gives you the following information: "One of these cabinets contains one million dollars. The other two are empty. I will ask you to choose one of the cabinets.

"Then, just to keep the audience entertained, I will open one of the other cabinets and reveal that it is empty. I will always choose an empty cabinet to open, and never your original choice. After that I will give you the opportunity either to stick with your original choice or to switch it to the remaining unopened cabinet. I will open whichever of the two cabinets you have finally chosen. If it contains the million dollars, the money is yours."

The game starts, and you choose the left-hand cabinet. The host opens the right-hand one and shows it is empty. The million-dollar question that confronts you is, 'Is it worth changing your choice to the middle cabinet? Or does it make no difference to your chance of winning?'

Most people (including physicists and mathematicians) reason incorrectly when they first meet this problem, along the following lines: 'The fact of whether the middle cabinet contains the money cannot have changed as a result of all this flim-flam. Therefore, there is no rational reason to change my choice. There are two unopened cabinets; there is an equal chance that the money is in either.'

But they are profoundly mistaken. Because although the physical situation has not changed, your ignorance has reduced—and that can make it quite rational to change your choice. Your ignorance about whether the money is in your original choice of the left-hand cabinet has not changed. It is still a one-third chance, as it was at the start of the game. But your ignorance about which of the other two cabinets has the money, assuming you originally guessed wrong, has disappeared. The chance that you originally guessed wrong is two-thirds, and in that case the money must be in the middle cabinet. You double your chances of winning by switching your choice. Thus a change in your knowledge of the universe—as happens when you make a measurement of a quantum system—can revise your expectation of the probable results you will get from subsequent measurements. To a naive person this might look as though acquiring knowledge about the system actually changed the system—like guests on the Monty Hall show discovering that changing their initial choice did indeed give them a two-thirds chance of winning, and then falsely thinking that this implied that money sometimes jumped from one cabinet to another as a result of their first measurement.

Born's approach was and is greatly respected. The rules of quantum probability are still widely referred to as Born rules. But as we have already seen, the most troubling observer effect, the EPR paradox as illustrated by the Bell-Aspect experiment and the lottery cards, cannot be explained by mere reduction of ignorance in a classical universe.

de Broglie and Einstein

Albert Einstein initially preferred an idea proposed in 1926 by Louis de Broglie. In this model the particles of radiation and matter are real and pointlike (or at any rate, very small) and their wavelike behavior is explained by their association with a kind of phantom field, which is detectable only through its effect on particles. This is, of course, the concept of guide waves. As we have seen, you can explain a great deal of what goes on in quantum by postulating some kind of invisible fine structuring to the world that can guide and jostle particles in a wave-like manner, describable by mathematicians in terms of hidden local variables.

However, Einstein soon came to realize the huge difficulties that nonlocality posed for this picture. In one of the most famous scientific papers of all time, written with Boris Podolsky and Nathan Rosen, he described the nub of the problem: After two particles have in some way interacted and traveled far apart, measuring one of them appears to have an instant effect on the other. The problem has been known ever since as the EPR paradox, or simply EPR.

Einstein hoped for a simple solution: Such long-range effects would turn out not to exist. Either quantum theory was incomplete and required modification or, more likely, there was some kind of error in the reasoning that implied that such "spooky forces" were operating. Einstein clearly thought that special relativity implied that no kind of influence could travel faster than light, irrespective of any quibbles about whether it could transmit information.

Nowadays, it is easy to borrow a glib psychologist's phrase and say that he was in denial but at the time it was a perfectly reasonable position to prefer the implications of special relativity, which had been thoroughly tested, to those of quantum theory. At the time he was pondering these matters in the 1930s (and even when he died in 1954), there was no practical way to investigate the matter experimentally. It was not until the 1960s that John Bell formulated the theoretical basis for an experiment that would be both definitive and practicable, and not until the 1980s that Alain Aspect and others were able to turn the experiment into foolproof hardware. But now the test has been done many times, and there is no question about it.[1] Nonlocality is real. Einstein was wrong.

Bohr

Niels Bohr is generally remembered as the father of the Copenhagen interpretation. Many textbooks describe the Copenhagen interpretation, formulated in dialogues between Bohr and others in and around that picturesque Danish city, as being the orthodox or mainstream interpretation of quantum mechanics. Yet there is no general agreement on what the Copenhagen interpretation actually is. At the lowest common denominator, it can be summed up in the following pair of statements:

1. The only real things are the results of experiments as measured by conscious, macroscopic observers; there is no deeper underlying reality.

2. Experiments yield results consistent with either wavelike behavior or particle-like behavior, depending on the design of the experiment, but never both at the same time.

But until the Copenhagen interpretation came along, the whole *point* of doing experiments was to formulate a picture of an underlying reality. Why, exactly, are we being forbidden to speculate further in this instance? Surely the idea that there are questions that must not be asked is contrary to the whole spirit of scientific endeavor.

Of course, there is nothing unreasonable about saying that a question is unanswerable because the result you get depends on the way that the question is asked. Consider, for example, a punchbag filled with a thixotropic fluid—one that acts like a liquid under gentle forces but like a solid if struck hard. The question "Are the contents of this bag liquid or solid?" can be answered only in the context of whether it is going to be squeezed or struck. But of course we can and do ask questions like: What is the threshold at which the behavior changes? Why does this happen? What, exactly, is going on at the molecular level? Bohr, by contrast, seemed to dismiss many questions about quantum as altogether meaningless, analogous to asking: "What color is up?"

Evidently Bohr felt confident that quantum theory as then formulated could answer all the questions that *he* felt it needful to ask of it. But he resisted further probing with wordy statements that have led many to retreat in confusion ever since, convinced that the tougher questions they had dared to pose had indeed been foolish. To me, Bohr's attitude seems uncomfortably reminiscent of those Buddhist sages who feel free to reply to certain questions with the response "Mu!" which means "The question is unsaid!" But many people have faith in such gurus.

My views on Bohr have recently undergone a partial change as a result of an intriguing paper by Don Howard, first delivered at a conference in Oxford.[2] Howard argues plausibly that the so-called unified Copenhagen interpretation was a myth invented retrospectively by Bohr's enemies (or at any rate, enemies of his school of thought), Heisenberg and Popper. In Howard's view, Bohr, far from being intentionally mystical in his replies, was merely being careful. If Howard is right, the nature of Bohr's caution is perfectly described by an anecdote many people will have heard in different forms. In my version, a child, a physicist, and a philosopher are traveling in a train passing through a country none of them has previously visited. The train passes a field in which they see a black sheep.

"Wow," says the child, "look at that. All of the sheep in this country are black!"

The physicist smiles. "We don't know that," he says. "All we can really tell is that some of the sheep in this country are black."

The philosopher smiles. "We don't know that," he says. "All we really know is that at least one sheep in this country appears black on at least one side!"

In an everyday context, we might consider that the physicist was the most sensible of the three. But if we are visiting a truly unfamiliar place—such as the world of quantum—then the philosopher's point that you should make statements only about the things you directly perceive, avoiding even the most reasonable-seeming inferences, is quite logical. Only by sticking to what you know for sure will you gain a reliable understanding.

When Bohr insisted that all it is legitimate to say about a quantum experiment is: "The experimenter observes such-and-such result," as opposed to "The quantum system was in such-and-such state," according to Howard he was merely being as careful as the philosopher in the train story. He certainly was not assigning any mystically powerful role to conscious observers. Interpreting any of his statements as "Conscious observers are the agents who physically trigger quantum collapse" would then be as much of a blunder as the famous mistranslation of the *canali* (channels) that the astronomer Schiaparelli thought he had seen on Mars as "canals," implying that Schiaparelli was postulating intelligent (and presumably conscious) Martians as the agents that created them.

I accept Howard's claim that Bohr was, at worst, a cautious agnostic, rather than a mystic. Possibly he hoped that if other investigators followed his example of making statements about only what they observed, rather than what they presumed, then a fully objective picture of quantum would ultimately emerge. But to me there seems a touch of cowardice about his stance. It was certainly frustrating to talk to Bohr; famously, he once reduced Heisenberg to tears. Here is an example of a genuine Bohr statement quoted by Howard:

> The quantum postulate implies that any observation of atomic phenomena will involve an interaction with the agency of observation not to be neglected. Accordingly, an independent reality in the ordinary physical sense can neither be ascribed to the [atomic] phenomena nor to the agencies of observation. . . .

> This situation has far-reaching consequences. On one hand, the definition of the state of a physical system, as ordinarily understood, claims the elimination of all external disturbances. But in that case, according to the quantum postulate, any observation will be impossible, and, above all, the concepts of space and time lose their immediate sense. On the other hand, if in order to make observation possible we permit certain interactions with suitable agencies of measurement, not belonging to the system, an unambiguous definition of the state of the system is naturally no longer possible, and there can be no question of causality in the ordinary sense of the word. The very nature of the quantum theory thus forces us to regard the space-time co-ordination and the claim of causality, the union of which characterizes the classical theories, as complementary but exclusive features of the description, symbolizing the idealization of observation and definition respectively.

If you find this less than transparent, you have my sympathy. It sounds rather deep. But try rereading the passage, changing the words "quantum" to "Olympian" and "atomic phenomena" to "gods," and you will see just how unsatisfactory the above statement is.

Bohr's answer to the specific problem of wave-particle duality is particularly inadequate. He said, essentially, no more than that we should expect a particle-like result from a particle-oriented experiment, and a wavelike result from a wave-oriented experiment. To me, this is uncomfortably suggestive of an engineer's rule-of-thumb. Imagine that you meet a hydraulics engineer who tells you the following story:

"I have two formulas that tell me exactly how fast water will flow through a channel of given size, under a given pressure difference," he says. "One formula works well for flow through narrow pipes, as used in domestic plumbing. The other formula works well for large constructions, like canals and aqueducts."

You take a look at his formulas. "But these are two completely different equations!" you exclaim. "They are supposedly describing the same thing, but would predict completely different results if they were applied to the same channel. What happens in a pipe of intermediate size, say one that is 10 centimeters in diameter?"

The engineer shrugs his shoulders. "I do only domestic plumbing and canals," he says cheerfully. "I don't need to know the answers for intermediate sizes."

Apart from his lack of theoretical curiosity, this hypothetical engineer would be missing out on his appreciation of a most important phenomenon: turbulence. The different equations arise because the flow through a narrow tube tends to be smooth or laminar, whereas larger flows naturally break up into the swirls and eddies of turbulence. Understanding turbulence is not only of great theoretical interest; manipulating the conditions that trigger its onset is the key to harnessing the properties of fluid flow in all sorts of contexts.

Nowadays, we can do experiments involving behavior that is intermediate between particle-like and wavelike. We are beginning to understand a process called decoherence, which is arguably the real mechanism of quantum collapse and is in some ways quite analogous to turbulence. Bohr has absolutely nothing to say about these kinds of situations. Agnosticism is perhaps an intellectually respectable position, but it does not lead to progress. Bohr had not so much an interpretation of quantum mechanics as an absence of one.

The worst part of the Copenhagen legacy, though, is that it continues to give aid and comfort to those who, in the debate between physicists and philosophers over the meaning of quantum theory, could be described as at the extreme philosophical end of the spectrum—those who maintain that questions about reality beyond the scope of immediate personal observation are meaningless. This solipsist viewpoint is impossible to refute, just like such claims as, "You have actually been lying on a couch all your life, wired up to a virtual reality machine," or "The world, complete with your memories and those of everybody else, has just been created in the last second." But it is utterly barren and unhelpful to the scientist's quest to build a meaningful picture of the universe. As Howard has pointed out, this idea has remained in the running largely because various claimants have muddied Bohr's name by falsely associating him with this viewpoint in a Copenhagen synthesis that never was.

von Neumann and Wigner

The mathematician John von Neumann's major contribution to the world was to lay the foundations for the computer revolution that followed later in the 20th century. But he also worked on the quantum theory, and his book, *Mathematical Foundations of Quantum Mechanics*, published in 1932, was fundamental to the field.

Von Neumann was the first person to think really deeply about the problem of quantum collapse. He was troubled by the potential for infinite regress, which we have already come across. If system A is measured by being put in contact with a larger system B, the result is measured by being put in contact with a still larger system C, and so forth, where does the process stop? When does the universe decide, OK, that's it, and settle down to a particular version of reality rather than tracing out yet more families of wavy variants? Von Neumann identified a physical need for collapse that goes beyond the philosophical problem of why we observe a single fixed version of reality. He realized that the equations of quantum are time symmetric. This, of course, contrasts with our macroscopic experience that there is a clearly defined arrow of time; eggs do not unscramble themselves, for example.

In the classical world, the arrow of time is associated with a steady increase of entropy, which can also be understood as a decrease of ordering. The universe started in an extremely ordered state, crammed into a tiny space, and even today most of its visible mass remains packed into stars occupying a very small fraction of its overall volume. The temperature difference between those stars and the cold emptiness of interstellar and intergalactic space provides the flow of energy that drives such processes as life on Earth.

But how does this tie in with the timeless quantum world, whose mathematical waves flow symmetrically without anything corresponding to an arrow of time? Von Neumann worked out that there is an entropy increase associated with quantum collapse, when multiple possibilities reduce to a single outcome. This is an interesting finding, but of course it requires physical collapse to occur at some point. Von Neumann reasoned that in the absence of any evidence for its happening earlier, the collapse should be assumed to take place at the point where a conscious observer inspects a quantum system.

To be fair to von Neumann, we must remember that he was writing before such basic thought experiments as Schrödinger's cat and EPR had even been formulated. I strongly suspect that he would have revised his views if he had lived until a later era. The contrast between his granting quantum collapse an important physical role on the one hand, and attributing it to an almost mystical cause on the other, is bizarre. But the idea of a conscious observer with a mysterious power to collapse systems by looking at them has appealed so strongly to a certain breed of thinker that it has survived for many decades. For example, von Neumann's ideas were still being extended in the 1960s by Eugene Wigner.

Wigner suggested that von Neumann's hypothesis from four decades earlier should be taken literally. Thus in Schrödinger's cat experiment, the point at which the cat's fate is determined comes not even when the box is opened, but when a conscious observer becomes aware of the result. For example, if the cat box is in space out beyond Pluto, aboard an unmanned probe with an automated opening mechanism that reveals the box's interior to a television camera, the cat's fate is not decided until the TV signal reaches the inner solar system. However, if an astronaut observer has been sent out to watch from aboard a nearby spacecraft, the cat's fate is decided as soon as he can see it, microseconds after the box opens. This is known as the paradox of Wigner's friend. (One wonders what he proposed doing to his enemies.)

Wigner's ideas have been rightly lampooned, by John Bell among others. Among the *reductio ad absurdum* questions one can ask are:

"What happens if there is a conscious observer in the box with the cat. Does the cat then die immediately, before the box is opened?"

"What exactly counts as a conscious observer? Is the cat a conscious observer? If so, what about a mouse, a frog, a slug? If not, what about a chimpanzee, or a Neanderthal? Where does the dividing line come? Does the observer need a PhD?"

The beautiful point has been made that in the context of cosmology, there were no conscious observers at all until a certain point (probably quite recent) in the universe's history. Was the entire universe waiting to collapse into a definite state until the first ape-man came along?

Conscious observers with spooky powers to collapse systems up to the size of a universe seem rather implausible. In any case the conscious-observer-collapse hypothesis does nothing to resolve the real problem of quantum, nonlocality. Remember the lottery cards example where my partner went to Mars. What happened when we simultaneously scratched our cards? Did my observation collapse the universe, or did hers, or was it both? In whichever case, the effect must presumably have rippled out faster than light to ensure that the far-off lottery card got to be the correct color.

Bohm

Other attempts to extend the early interpretations of quantum were more respectable. Perhaps the most heroic attempt to cling to a classical picture is found in the rather tragic story of David Bohm, an American physicist who came to England's Birkbeck College in London when his services were no longer required on the Manhattan project.

(I cannot resist pointing out here a curious fact about the scientists who have contributed the most to our understanding of quantum theory: A remarkably high proportion have four-letter surnames beginning with "B." Those we shall encounter include Born, Bohr, Bohm, and Bell; there was also Bose of Bose-Einstein condensate fame. And some claim there is nothing weird about the statistics of quantum!)

In the 1950s, Bohm effectively rediscovered and revitalized the pilot-wave theory which had been invented by de Broglie a quarter of a century earlier, but fallen out of favor because of its problems with nonlocality. Bohm was determined to make the pilot-wave theory work somehow, despite the apparent faster-than-light influences of EPR. Mathematically, his work was to an extent successful and his findings interesting. He discovered that pilot-wave theory could work after a fashion if you assumed that the guide waves continued indefinitely even when they were no longer associated with any particular particle. He found, however, that individual guide waves did not, as you might expect, die away with time, viewed from a macroscopic scale, like water waves decaying to ripples after the storm that caused them has died down. The guide waves can remain large in amplitude even at times and places remote from their last occupation by a surfing particle. One way to see intuitively why this must be is to reflect that warships carry charts of equal scale and detail covering every portion of the world's oceans—because although there are places they are most unlikely to be ordered to, if they ever are, then the scale of maps required to navigate properly is just the same as for those regions they visit frequently. Pilot waves, if they do exist, must guide particles with accuracy through low-probability as well as high-probability regions.

(Ironically, as David Deutsch and others have pointed out, Bohm's work is excellently supportive of many-worlds. If you forget the rather artificial notion that the waves are occupied by surfers whose positions define a single reality, then the waves are tracing out all possible world-lines with equal fidelity. But we shall come to this later.)

Bohmian mechanics, as it is now called, is more sophisticated than the simple surfing-particle story we constructed in the previous chapters. Guided by an extra field he called the quantum potential, his particles did not "tumble off" their guide waves on undergoing measurement-like interactions; rather, the wave that they were riding underwent a subtle and pseudo-instantaneous change.

But of course this quantum potential has to operate in a nonlocal manner, and Bohm's attempts to explain how this could happen became rather desperate. In a book written with Basil Hiley shortly before Bohm's death (it was published posthumously), he introduces the notion of implicate order.[3] The attempt to understand this concept has baffled many physicists, but I think the idea can be taken in two separate ways. We can understand it as saying either that there is a kind of limited but instantaneous linkage between all parts of the universe, which is not directly accessible from the macroscopic world where time's arrow operates, or that the universe contains embedded in every part of itself encoded information about what is going on in the other parts.

The problems with the first view we have already examined. They include contradictions with the spirit of special relativity, implying backward-in-time causation at the micro scale, as well as the truly enormous amount of behind-the-scenes calculation that the universe must be assumed to do if every part of it can instantly affect every other part. The problem with the second view is that it implies predestination. If a billion projectors are playing exactly the same movie, without needing to communicate with one another, then even if each projector is showing only one particular area of each frame at maximum magnification, they must surely all have been loaded with the same film at the start of the performance.

Price, Valentini, and Cramer

There is nothing intrinsically impossible about predestination. Richard Feynman was struck by the symmetry between the processes by which radiation is absorbed and emitted. We normally think of radiation as going out from its source in all directions, but approaching a target from only one direction; yet it is just as reasonable to think of it converging on its target from all directions, an invisible noose drawn tight with perfect precision. Feynman and others have toyed with the idea that although, to our usual perception, the order of the universe is decreasing, perhaps in subtle ways it is increasing. The universe originated in one rather special highly ordered state, and is progressing toward another, rather than toward pure disorder. From our point of view, this progress toward a future constraint looks exactly like predestination. As Professor Cope explained in Chapter 1, predestination can easily account for EPR correlations. Indeed, predestination can in principle explain *any* apparently nonlocal phenomenon. Think of two people a light-year apart, holding a conversation; if each knows exactly what the other is about to say and when, each can "react" to the other without any delay for a signal to go from one to the other. The idea that a specific, subtle kind of predestination can explain EPR and other puzzles of modern physics has been developed by Huw Price, and is described in his popular book *Time's Arrow and Archimedes' Point*.[4] Unfortunately, Price has encountered considerable technical difficulties in trying to develop a predestination theory that takes account of the way subatomic particles are known to behave.[5]

But the real trouble with the postulates of predestination and instant everywhere-to-everywhere links is that they are much *too* powerful merely to explain EPR correlations. If such phenomena exist, the problem becomes: How does the universe implement such remarkably efficient *prevention* of apparent faster-than-light causal effects and faster-than-light communication of information? Where does the censorship come from? Italian physicist Antony Valentini has attempted to develop a kind of hidden-variable theory in which the early universe did have general faster-than-light causal links, which died away naturally, except at the microscopic level, due to thermodynamic considerations, but his views have not won wide acceptance. Valentini has been brave enough to suggest an experiment to test his ideas. Essentially, the idea is to use a powerful telescope to capture photons that were emitted very early in the history of the universe, and subject them to a two-slit experiment. He predicts that the usual interference pattern will not be found. I applaud his courage, but I (and many others) would be prepared to bet a substantial sum that no such anomalies will be found.

Another variation on the predestination theme is John Cramer's transactional interpretation. Cramer invites us to imagine a "retarded wave" spreading backward in time from the point at which a system is finally measured, for example, by the absorption of a photon in a specific spot, and interfering with the forward wave which we normally think of as constituting or guiding the photon. You could think in terms of a sort of "negotiation" or transactional discussion between the past and the future that decides whether, for example, Schrödinger's cat lives or dies. But this way of thinking has proved too cumbersome to gain widespread acceptance.

Conclusion: Whither?

The work of Price, Valentini, and Cramer actually represents the respectable end of an endeavor that has become increasingly unrewarding, trying to cling to quantum interpretations invented a lifetime ago. Other attempts have led even powerful intellects into dubious pathways. I recently heard a rather distressing talk by a former colleague of Bohm's detailing how his immediate circle at Birkbeck developed some aspects of a cabal, complete with initiation rites, as an ever more isolated group attempted to explain away the contradictions of quantum with ideas borrowed from literary theory and even psychoanalysis. Bohm died, in the judgment of many who knew him, "badly bewitched by philosophy." Philosophical discourse into quantum has taken some unhelpful turns, most especially with respect to the claim that asking certain questions about quantum systems is meaningless or forbidden. Because everything in the universe is in fact a quantum system, an extension of this attitude could pretty much spell the end of scientific endeavor.

We will eschew philosophical excuses and outdated notions. We are looking for a physical, visualizable solution that our common sense can accept. Because none of the ideas above properly address the PPQs we have formulated, we must look for newer ones.

CHAPTER 6

LET'S ALL MOVE INTO HILBERT SPACE

There is one way in which quantum mechanics has indisputably progressed since the interpretations discussed in Chapter 5 were invented. To understand it, we must prepare to visit a rather strange place that was invented by the mathematician David Hilbert. It is called Hilbert space in his honor.

State Space

The key idea is that in a space with a sufficient number of dimensions, a single point can describe the state of an entire system, however large. We'll start with a simple example. Let's suppose you own a trucking business that transports goods between New York and Chicago. If you own just one truck and it is always somewhere on the interstate highway between the two cities, you can indicate its position at any given moment by a point on a one-dimensional graph, a straight line, as in Figure 6-1a. The truck is driven by Albert.

But now let's suppose that your business expands to two vehicles, with a second truck driven by Betty. You could indicate their positions using two different points on your original graph, by using two different markers. But you could also indicate their positions using a single point on a two-dimensional graph, as shown in Figure 6-1b, where the horizontal axis gives Albert's position and the vertical axis Betty's position. If you got a third vehicle and driver, you would need a three-dimensional graph to keep track of the whole fleet with a single point, as in Figure 6-1c, and so on. Obviously, you will need an *n*-dimensional graph to keep track of *n* trucks. You cannot readily visualize a graph of more than three dimensions, of course, but it is perfectly possible to handle mathematically.

If you switch your business to operating a fleet of ships, you will need a graph with two dimensions for each ship, because ships are not confined to roads, and can freely roam a two-dimensional surface; it takes two coordinates per ship to record the latitude and longitude. Aircraft would need three coordinates per vehicle, to include the altitude. To know what orbit a spaceship is going to follow, you need to know not only its position but also its speed in the x, y, and z directions, so it takes a graph of six dimensions to record the full trajectory information for one spaceship. If you have 10 spaceships, your graph needs 60 dimensions, but a single point on it still records all the information about your fleet that you need to know.

Of course what we're really interested in is not trucks or spaceships but fundamental particles. If the universe consisted of pointlike classical particles, we would need $6N$ dimensions to keep track of a system of N particles including their positions and speeds: 12 dimensions for a two-particle system, 18 for a three-particle system, and so on. There are perhaps 10^{80} particles in the observable universe, so a single point in a space of about 10^{81} dimensions could record the exact state of the entire classical universe. [1] If that sounds like a lot, just wait. . . .

Probability State Space

Quantum systems are more complex than classical ones and require more information to describe them. Suppose you are back to owning just one truck, but it's a quantum one. Even if it sticks to the route between New York and Chicago, its position is described not by a dot on a line but by some kind of probability wave having a specific value

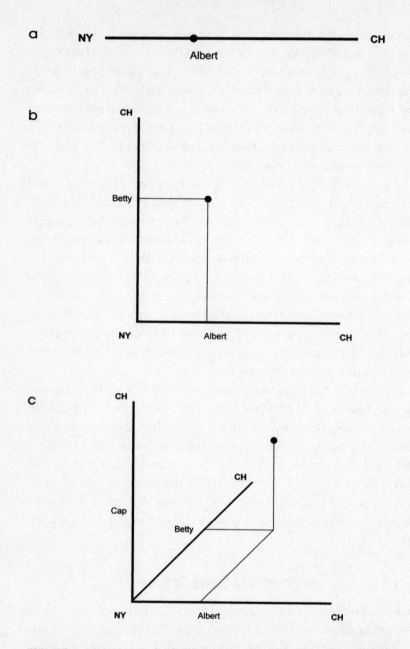

FIGURE 6-1 Keeping track of (a) one truck, (b) two trucks, (c) three trucks.

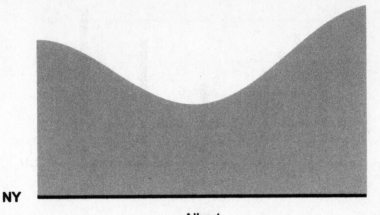

NY   CH

Albert

FIGURE 6-2 Albert's probability wave.

at every point along the route, as shown in Figure 6-2. The shape implies that Albert tends to loiter near the ends of his route.

This is bad news for our project to record all the information about his position in the most compact way possible. To fully record the information in Figure 6-2, we would have to write down the height of the graph at every point along the x axis; an infinity of points, so an infinity of values. Things get more manageable if we only need to know roughly where Albert is: say, in which county out of 5 counties along the route. Then we get a bar chart as shown in Figure 6-3a. The information is given in the height of 5 individual bars, 5 numerical values, and we could record it by placing a dot at an appropriate position in a space of 5 dimensions. If we add a second truck, driven by Betty, her graph might look like that in Figure 6-3b, and we could record the position information for both trucks by a point in a 10-dimensional space. This is worse—in the sense of more extravagant—than the situation for classical trucks or particles, but not that much worse, for the basic rule is still *additive*. A two-particle system will require a space of twice as many dimensions to describe it as a one-particle system.

For display purposes, you could combine the information on both the graphs into a single 3-dimensional graph, as in Figure 6-3c. How-

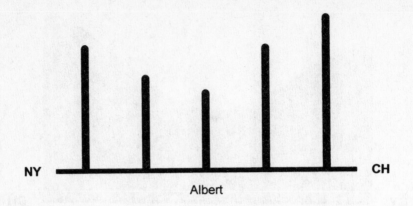

FIGURE 6-3a Probability of Albert being found in each location.

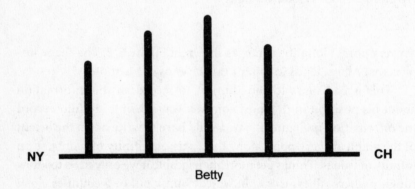

FIGURE 6-3b Probability of Betty being found in each location.

ever, at the moment, Figure 6-3c does not contain any more information than 6-3a plus 6-3b. Although it has 25 columns, a point in 10-dimensional space still contains all the information we need to generate it.

But now let's introduce Albert to Betty. The results are dramatic. They start to interact; indeed, they fall in love and get married. It's all very sweet, but now the interaction makes the probability wave describing where your trucks are much more complicated, as shown in Figure 6-3d. For example, in many places the probability that Albert

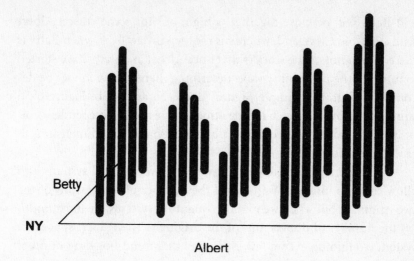
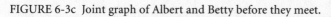

FIGURE 6-3c Joint graph of Albert and Betty before they meet.

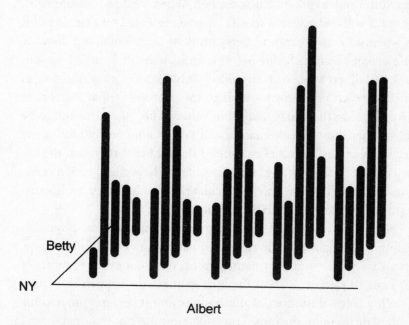
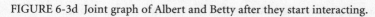

FIGURE 6-3d Joint graph of Albert and Betty after they start interacting.

and Betty will be close together is high, but for some reason Albert tends to avoid Cleveland, where his mother-in-law lives, when Betty is there. The point of the story is that once Albert and Betty have started to interact, the probability wave describing them can no longer be decomposed into two simple Figures like 6-3a and 6-3b: Figure 6-3d simply contains too much information. There are 25 independent columns, so it would now require a point in a space of 25 dimensions to record the information.

The general rule is that when we join two classical systems and allow them to interact, we just add the number of dimensions of the two originals; but when we join two quantum systems, we must *multiply* the number of dimensions of the originals. The Hilbert space of a system containing even a few particles has a mind-boggling number of dimensions.

What is the significance of Hilbert space? Hilbert space represents a quantum system before it is measured. When we do a measurement, the space will collapse to a specific state represented by a single point (as when we ring Albert and Betty on their cell phones and discover their actual positions). But before a measurement is made, we can think of Hilbert space as being filled with a kind of grey mist whose density at each point corresponds to the probability that the system will collapse to that particular set of values. This mist turns out to be highly amenable to mathematical analysis: it slops around following very simple rules, in fact even simpler than those that govern the behavior of a real fluid like water. So, despite the large number of dimensions involved, the best way to calculate the evolution of an isolated quantum system is to use Hilbert space.

Now you understand the dilemma of mathematical physicists dealing with quantum. The rules of quantum are beautifully simple. But all except the simplest quantum processes—for example, those of tiny isolated systems such as the hydrogen atom—happen in a space of such a colossal number of dimensions that it becomes impossible to simulate them on the most powerful computers now available, and utterly hopeless to try to visualize them with our own minds.

A Space of Her Own

Hilbert space is usually described as totally abstract, utterly remote from the three-dimensional world of our ordinary perceptions. But there is a well-known experiment that calls for creating a large bubble of Hilbert space, embedded within the everyday world, which you could in principle touch. We have encountered it already. I am talking about Schrödinger's famous cat.

To really do Schrödinger's famous cat experiment, we would need to create a cat box that no information could leak out of, a box of macroscopic size that was truly and utterly sealed from the outside world. Just for fun, let us see if we could conceivably do this with present or near-future technology. In the interests of both scientific progress and cat welfare, we will replace the cat with a human ob-server, a kind of philosopher-astronaut.

It is vital that the box does not touch anything, so we will start by going into space where it can be allowed to float free in microgravity without any connecting struts. To shield against the high-energy charged particles called cosmic rays, which are found everywhere in space, we will hollow out a chamber at the center of some natural object, an asteroid or comet nucleus. We must then protect the central chamber against particles caused by radioactive decay of elements in the asteroid, probably with a thick shield of pure metal. I cannot resist telling you an odd fact at this point. Since the first atomic tests in the 1940s, all the steel made on Earth has been slightly contaminated with radioactive particles present in the atmosphere, which inevitably get into the blast furnace because large amounts of air are needed for com-bustion. When scientists need steel that is completely free of decaying radioactive nuclei, they get it from a surprising source. After World War I, the German battleship fleet was scuttled at Scapa Flow in Scot-land, the giant natural harbor where the British Grand Fleet used to be based. When nonradioactive steel is required, scuba divers go down and carve chunks of pre-Atomic-age steel from the battleships, which thankfully provide a huge resource of the material. Within the aster-oid, we will construct a thick sphere of this ultrapure steel.

The cat box, or philosopher box as it is now, floats within this central shield, and within it is a further thin spherical shell like a Christmas bauble that is cooled to the lowest temperature practicable, in the milli- or perhaps even micro-Kelvin range, to suppress the radiation of infrared photons. The inner shell contains a perfect vacuum except for the very occasional very low energy infrared photon emitted by the walls. Because such photons have a wavelength of several meters, they reveal nothing about the position or state of the central chamber beyond the fact that it is there.

In theory, the space capsule now no longer contains a philosopher-astronaut, but a Hilbert space, a probability distribution of philosopher-astronauts doing increasingly divergent things, as their personal histories diverge depending on exactly how many photons hit each cell of their retinas and other quantum events that multiply into macroscopic consequences in various ways. If we could look inside the capsule (which is, by definition, impossible), we might imagine seeing something like a multiple-exposure photograph. Is the astronaut writing, or brushing her teeth, or just staring into space? This is the image that inspired the title of this book. Rabbits are famous for their tendency to multiply; what a Schrödinger box really contains is not one of what we originally put in it, but many.

We have seemingly created a macroscopic bubble of Hilbert space, in which different probability histories of the astronaut, eventually diverging quite significantly, can trace themselves out. In principle, we could do a test to prove that this has happened, using interference between the different histories, and we will return to this possibility in the last chapter. However, when the capsule is opened the astronaut herself will report nothing out of the ordinary—the Hilbert space will instantly collapse to a single point, selecting just one of all the possible states that it has been exploring.

Alas, there is at least one effect that might still make this experiment impossible, despite all our elaborate precautions. One effect that we do not know how to shield against is gravity. Although the center of the capsule will automatically remain in the same position no matter how the astronaut moves about, only a perfectly symmetrical object can have a perfectly spherical gravitational field. A real object—like Earth or a space capsule with an astronaut in it—has a field with subtle variations betraying information about the internal disposition of its mass. Remember Borel's thought experiment in which shifting a small rock light-years away could change the positions of air molecules in Earth's atmosphere, via gravitational effects amplified with every molecular collision? It is difficult to calculate the extent to which such effects would continuously measure the capsule in the above experiment, but it might well be sufficient to make the macroscopic superposition we are trying for impossible.

Natural Collapse

The analysis of Hilbert space has thrown an extraordinary new light on the process we call quantum collapse. In 1970, Dieter Zeh at the University of Heidelberg demonstrated something remarkable. In a system that evolves in Hilbert space, whose components interact significantly, the mathematics predicts that although at first sight things appear to proceed quite unselectively—there is no telling, for example, what position one particular particle is likely to occupy—patterns nevertheless start to emerge that are durable in the sense that they continue to be strongly affected by patterns of high co-probability, but in a rapidly decreasing fashion by patterns of low co-probability. The mathematical process by which inconsistent patterns exert increasingly small effects on one another is called decoherence.

Decoherence can effectively explain quantum collapse—or at least apparent quantum collapse. To see how, let us consider a nested system of Schrödinger's cats. Assume that the astronaut described above takes into the capsule with her a small cat box designed on the same lines, with Schrödinger's original diabolical arrangement that might kill the cat with 50 percent probability.

We seal the capsule. From our point of view, both the astronaut and the cat are in Hilbert space. But we know that after a certain time, she will open the box. What does the Hilbert space model now reveal? It tells us that as soon as she starts to open the cat box, the possible states of herself very rapidly become entangled with those of the cat. There are states of her that are rejoicing, having found a live cat, and states of her that are mourning, having found a dead one. But these different states very rapidly cease to have a significant effect on one another. Each state of her has apparently seen a quantum collapse in which the cat has become definitely dead, or definitely alive.

At this point it is really impossible to avoid a mention of many-worlds because: What happens when you open the astronaut's capsule? You are going to see either a happy woman with a living cat in her arms, or a sad woman holding a dead one. If you accept that the Hilbert space analysis applies to the whole universe, then what is really happening is that one version of you is becoming correlated with the happy-live-cat outcome, and another version of you with the sad-dead-cat one.

We will have more to say about this later, but I am certainly not yet claiming that this is a proof of many-worlds. A single-worlder might describe the opening of a Schrödinger's cat box something like this:

"When I opened the box, the outside environment started to measure what was in there. The very first measurement photon out of the box might give a strong clue—for example, if it was an infrared photon at the temperature of a live cat.

"Just as when you scratched one lottery card, it made a certain outcome of scratching the other more likely, so a measurement consistent with (say) a living cat makes subsequent measurements consistent with that outcome more likely. And so either a live cat or a dead one emerges, rather than some gruesome combination. From the abstract processes of Hilbert space, consecutive measurements brought a specific consistent reality into being."

The single-worlder might have a point. Despite my simplified account above, it remains controversial whether you can in fact get sensible numbers out of Hilbert space without some form of context dependence—some privileged starting point such as a unique reality from which you can measure everything. But we will postpone this argument to a later chapter, and concentrate for now on the solid achievements of decoherence.

Testing Decoherence

Decoherence theory allows us to calculate exactly the timescale over which any given system will decohere—in the old language, the time for quantum collapse to happen. I am not going to describe the math, but it is useful to get some idea of how long collapse is predicted to take in certain situations. One sort involves the spatial localization of small objects whose position is measured from time to time by interactions in which they scatter photons and other particles in their vicinity. Table 6-1 is adapted from a paper by Erich Joos.[2]

The top left figure in this table tells you, for example, that a particle of dust a hundredth of a millimeter across (just big enough to be visible with a strong magnifying glass) that is floating in interstellar space, and whose position has become uncertain by a centimeter, is likely to pop to a relatively precise location in about a microsecond. However, if its position is uncertain to only about the same distance as its own diameter, a hundredth of a millimeter, it will take a second or so to get relocalized. Note the huge variation from the top right to the bottom left of the table. Relocalization becomes much faster as you approach Earthlike conditions of temperature and atmospheric pressure. It also gets much faster for larger objects. There is probably nowhere in the natural universe where objects larger than dust grains are delocalized to any significant degree, because the famous 3° Kelvin

TABLE 6-1 Localization Time (seconds-cm^2)

	$a = 10^{-3}$ cm Dust Particle	$a = 10^{-5}$ cm Dust Particle	$a = 10^{-6}$ cm Large Molecule
Cosmic background radiation	10^{-6}	10^{6}	10^{12}
300 K photons	10^{-19}	10^{-12}	10^{-6}
Sunlight (on Earth)	10^{-21}	10^{-17}	10^{-13}
Laboratory vacuum (10^3 particles/cm^3)	10^{-23}	10^{-19}	10^{-17}
Air molecules (standard atmosphere)	10^{-36}	10^{-32}	10^{-30}

cosmic microwave background radiation, remnant of the Big Bang, is all-pervasive.

There are subtler forms of decoherence than simple localization, however. Another system of interest is a regular oscillator whose motion is slowly decaying, like a swinging pendulum subject to friction. It turns out that the decoherence time of such a system is directly related to the damping time—that is, the time it takes for the pendulum's swing to decrease to half its original value. This link between quantum decoherence and the increase of classical entropy, the slowing of things due to friction, is a tremendously important theoretical result. Unfortunately for anyone hoping to witness a pendulum in a superposition of different angles of its swing (an effect you can sometimes see in trick photographs), the ratio of the decoherence time to the decay time is extremely small, of the order of 10^{40} in the case of a 1-gram pendulum on Earth.

However, this ratio is proportional to the absolute temperature of the surroundings, and to the mass of the object. It gets more reasonable for a small object spinning in a vacuum, an object for which the damping time is also extremely long, because there are only tiny effects tending to slow the spin. Such an object can remain in a superposition of different angular positions for an appreciable time, but again, naturally occurring examples are spinning dust particles in interstellar space rather than large terrestrial objects.

Feasible Experiments

Let us now return to Earth, however, and emphasize that decoherence is not just a theory. It can be tested in doable experiments. Such tests have already been performed by the redoubtable experimenter, Anton Zeilinger of the University of Vienna, with interference experiments using relatively large objects—fullerenes, football-shaped molecules whose basic form is a cage of 60 carbon atoms.

Zeilinger looked at ways in which environmental decoherence—that is, the environment "reading" the position of the molecules—tends to degrade the interference pattern obtained in a two-slit experiment. One such test involved doing the experiment in a space that was not a perfect vacuum, so that occasional collisions with gas molecules caused decoherence, degrading the interference pattern. Another used molecules that were hot enough to emit infrared photons as they flew along their trajectories, giving away information about their positions.

In both experiments, the predictions of decoherence were confirmed. The error bars were relatively large, but new experiments now being proposed should dramatically increase the accuracy. Indeed, as we'll see in a later chapter, devices like quantum computers, which are extremely sensitive to the effects of decoherence, naturally provide a way of measuring it to very high accuracy, and that is part of the motivation for trying to build such devices.

Unless something very unexpected emerges, the mystery of where, when, and how quantum collapse occurs must be considered solved. It is, quite simply, decoherence that does it. Dieter Zeh's hypothesis has been confirmed by 30 years of calculation and experiment, and it is something of an indictment of the system by which scientific advance is recognized and popularized that this tremendous progress is not better known.

In Quest of the Finite

There was one point about Hilbert space that I rather skated over: the fact that strictly speaking, the probability wave associated with even a single particle needs a Hilbert space of infinite dimensions to describe its exact value everywhere. If you have a good physicist's distaste for infinities, let me throw you a lifeline—in fact, two lifelines.

First, there is nowadays strong evidence from the field of general relativity that the maximum amount of information that can be stored in and retrieved from a finite-volume region of our three-dimensional universe is itself finite. There is even a formula for calculating it, called the Bekenstein limit after its discoverer. No one knows yet quite what implications this has for Hilbert space descriptions of the universe. There have always been awkward clashes between general relativity and quantum theory. But it is possible that it means that the number of dimensions required for Hilbert space is not quite infinite, merely mind-bogglingly colossal (still *much* larger than the mere 10^{81} dimensions or so required to describe a classical universe). So, wherever I have used the word "infinite" in connection with Hilbert space dimensions, you can possibly substitute "vast." The implications might be important, but this is a very controversial area that we will return to in the final chapter.

Be that as it may, there is one aspect of most kinds of particles that requires not an infinity, or even a mind-boggling number, of dimensions of Hilbert space to describe it, but exactly two. As well as having a position, many particles have a much simpler intrinsic property, called spin in the case of electrons and polarization in the case of photons. Spin and polarization represent *internal* properties of particles. In terms of our trucker analogy, they might represent something like the angle at which the driver's cigarette is currently pointing. (I regret to tell you that Albert and Betty are both chain-smokers.)

External quantum properties like position along the x axis must be represented by a waveform containing an infinity of real numbers, giving the probability of the particle being found at each possible point along the axis. After collapse, the result of a measurement is a single real number which still requires an infinity of digits to record its exact value, like 119.3564218. . . . The universe "knows" an infinity of real numbers, and gives you one back. But a quantum value such as polarization can be represented by just two real numbers—like the direction in which Albert's cigarette is currently pointing, given in terms of compass bearing and elevation—and when you collapse it, you get back *just one single binary digit*, a yes-or-no answer, as if all you can record from outside the truck is whether Albert ultimately discards the cigarette stub out the right- or left-side window.

The two real numbers describing spin could be drawn rather unimaginatively in a bar-chart with only two columns, but a neater way is shown in Figure 6-4, called the Bloch sphere after its inventor. Here the direction of the particle's spin axis (the direction Albert's cigarette is pointing) is shown as the latitude and longitude of a point on an imaginary sphere. On measurement, the vector shoots to either the north or south pole.

It was David Bohm who first realized that, in a less-is-more kind of way, using these modest internal quantum properties might yield

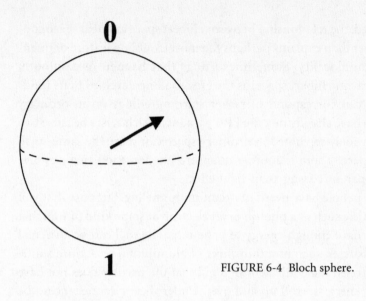

FIGURE 6-4 Bloch sphere.

the most practical way to perform the kind of test of quantum non-locality that we met in the Bell-Aspect experiment, far more doable than the experiment originally proposed by Einstein. Later, David Deutsch and others realized that harnessing these internal quantum values is also the way to a viable quantum computer. Before measurement, the uncollapsed vector shown in Figure 6-4 can be thought of as a qubit, the quantum equivalent of a binary digit; when it is collapsed by measurement, it becomes an ordinary bit, by assigning the values 0 and 1 to the north and south poles in the diagram. For more on quantum computers, see Chapter 11.

Not a Panacea

The arena of Hilbert space, and the process of decoherence, have given us deep insights into quantum that the pioneers who invented the old interpretations did not have. But the new concepts do not by themselves answer the key interpretational puzzles of our PPQs and they do not help us to visualize the processes of quantum in terms of the space and time we are familiar with, to tell ourselves a meaningful story of what is going on.

Indeed, the relationship between Hilbert space and real space conceals rather than explains perhaps the most troubling feature of quantum, its nonlocality. Sometimes things that happen in a smooth, orderly way in Hilbert space, as the gray shading develops in its fluid-like way, can correspond to rather startling goings-on in ordinary space. We have already met the EPR paradox, which arises because two photons widely separated in ordinary space can share the same small Hilbert space. There is another quantum phenomenon that can give rise to apparently faster-than-light effects.

Most people have heard of quantum tunneling. Suppose that you fire a particle such as a photon or an electron at some kind of wall that it doesn't have enough energy to penetrate. The wall can be an actual physical barrier, such as a thin sheet of aluminium, or a more subtle energy barrier, the equivalent of a hill that the particle does not have sufficient energy to roll up and over. Under these circumstances, the rules of quantum mechanics—specifically, the Heisenberg uncertainty principle—predict that because the probability wave associated with the particle slops over beyond the wall, occasionally the particle will appear to tunnel straight through what would otherwise be an impassable obstacle, just by happening to jump from one part of its probability wave to another.

A disturbing feature of quantum tunneling is that it appears to happen instantly. We can describe this in words: "Any time that you choose to measure the particle, you will find that it is on one side of the wall or the other. The implication is that at some point, it must have moved across the wall in no time at all."

Those are just words. But the mathematics does also seem to describe the particle leaping across the wall instantly, and when you watch a computer simulation of the probability wave associated with the particle, the central part describing where the particle is most likely to be detected does indeed seem to proceed faster than light. The matter is unclear enough that experimental physicists, including highly respected ones, have run tests to see whether they can transmit data faster than light using quantum tunneling, and some even claim to have succeeded.

Now, before anyone gets too excited, I hasten to add that you almost certainly can't do this. More careful work, both in computer simulations and with actual photons, indicates that although the probability waves do seem to travel faster than light at certain points, you cannot propagate a disturbance along them at this speed. You cannot really send information faster than light, with the paradoxes that could imply. My point is that a good visualization or interpretation of quantum should not even tempt us to think such a thing could happen. Hilbert space is a good place to do math, but it does not provide us with a clear intuitive picture of what is going on in the three-dimensional world.

CHAPTER 7

PICK YOUR OWN UNIVERSE

This chapter contains both good news and a warning. The good news is that there will certainly be more than one fully correct way to look at quantum. In a sense, we have an infinity of choices. The warning is that it is perilously easy to make a choice that hinders progress rather than helping it, and we shall look at some pertinent cautionary tales.

A Choice of Games

There is an ancient idea that events here on Earth are merely the actualization of a game being played between gods. It appears in Greek legends, in the Norse sagas of the Vikings, and in folk tales from cultures all around the world. Most people no longer believe in a panoply of gods who can control human beings like pieces on a board, but in the past century the metaphor was revived by the great physicist Richard Feynman. He wrote:

> We can imagine that the complicated array of moving things which constitutes "the world" is something like a great chess game being played by the gods, and we are observers of the game. We do not know what the rules of the game are; all we are allowed to do is to watch the playing. Of

course, if we watch long enough, we may eventually catch on to a few of the rules. The rules of the game are what we mean by fundamental physics.

As usual with Feynman's insights, this opens up a rich vein of thought, going far beyond the immediate purpose for which he used the metaphor. One such development is remarkably empowering when it comes to interpreting quantum physics.

Many readers will have heard of game theory, a field of mathematics whose applications include finding optimal strategies in such fields as business, military conflict, and deterrence, where the decisions of others must be taken into account. Rather more obscure is games theory, which concerns itself with real-life games such as chess and bridge—usually games involving boards, cards, dice, and suchlike accessories. Yet games theory, too, has practical applications. For example, a lateral-thinking approach to proving a mathematical theorem is this. Imagine a game played between two mathematicians: A, who wants to prove the theorem, and B, who wants to refute it. If we can prove that A has a guaranteed winning strategy for the game, then the theorem is proved, without needing to work through the details of all the various moves A and B can make.

An important insight from games theory—indeed, the foundation on which the whole field is based—is the fact that many games that appear quite different are, in fact, algorithmically the same game. A trivial example is the different editions of the game Monopoly that are played in different countries. In each country the properties are labeled after districts of a great city. Thus in the American edition, the most expensive property is called Boardwalk; in the English, it is Mayfair, once London's most fashionable area; in the French, it is the Rue de la Paix. But the rents and prices of the property remain the same in each case, so of course this relabeling makes no difference to the game. Nor does the fact that in the English version the prices are nominally in pounds, and the French in euros. In the old French version the prices were in francs, and because a dollar is worth about 10 francs, the numerical prices were all multiplied by exactly 10 with respect to the American version. The addition of an extra zero to all the currency bills and prices likewise had no effect on the play.

In the case of Monopoly, it is not hard for adults to distinguish the algorithmic part of the game from the story component. When you get a card informing you that you have won a beauty contest, or that it is your birthday, the significant information is of course how many points (money) are to be transferred from whose account to whose. But children can have difficulty even at this level. A few years ago, on a long airplane flight, I found myself sitting next to a charming lady whose son was happily occupied with a book on the game Pokemon, then a worldwide fad among preteens. I chatted with the mother and we agreed that the game separated the world sharply by age group. Whereas children were obsessed with it, almost no adults—even those who were parents of young children—had any idea of even the basic rules or objective of the game. The kid overheard our conversation, and decided that it was his duty to educate us, with amusing but un-productive results. He was trying his best to describe how the game worked, but could think of no other way to start than with the story preamble, telling us how the Pokemon characters (represented by cards) were mouse-sized creatures carried in bottles on a special belt. When it came to the cards themselves, realizing that a grown-up ap-proach was called for, he skipped embarrassedly past childish-looking creatures until he found something that fit the bill. "You'll like this one," he said proudly. "It's the Great Green Brain-Blaster, it's really good. . . ."

Desperate though he was to describe the essence of the game, he found it impossible to reach the required level of abstraction: "What really matters is the points number on each corner of the card. You add the ones in the top left-hand corners together, then subtract. . . ." Before we laugh too hard, we should reflect that adults, too, can find the distinction between story and essence harder than it seems.

⚭⚭⚭

Let us embark on a field trip. The idea is simple: to find some gods playing a game, observe them for a while, and figure out what the game is. A couple of thousand years ago, we would have climbed Mount Olympus but this being the 21st century, we will fly off in a spaceship until we discover some promising-looking gods, as shown in Figure 7-1.

We realize that the game these alien gods are playing might in fact

FIGURE 7-1 These aliens are playing a game that involves speaking words in alternation. By patient observation we discover that nine different words are used, and each is used a maximum of once per game. The players take turns to utter a word until the last one to speak wins. Games are a minimum of five and a maximum of nine words long (because all the permissible syllables have then been used). Nine-syllable games sometimes end in a draw with neither player winning, although shorter ones never do. We see that the rules are consistent. If a given sequence of syllables wins on one occasion, it does so on any subsequent occasion.

be quite simple, but unfortunately they are playing it in their heads, without use of board or counters. Yet it is vital that we learn the rules because in due course we want to be able to play against these gods and beat them. We might start by tabulating all the different possible games, as shown in Table 7-1, in which the first player's moves are in

TABLE 7-1 List of All Possible Games

XIG flump WIBBLE nias AG choo GAH	god 1 wins
XIG flump WIBBLE nias AG choo DOH gah FIZZ	god 1 wins
XIG flump WIBBLE nias AG choo DOH fizz GAH	god 1 wins
XIG flump WIBBLE nias AG choo FIZZ	god 1 wins
. . .	
XIG flump WIBBLE nias AG gah CHOO fizz DOH	drawn
. . .	
XIG flump WIBBLE nias GAH ag DOH choo . . . and so on.	god 2 wins

capitals and the second's in lowercase. Unfortunately the table will be rather long; it could have up to nine-factorial entries, that is, $9 \times 8 \times 7 \times 6 \times 5 \times 4 \times 3 \times 2 \times 1 = 362{,}880$. In practice it will be less than that, because many games end before a full nine moves have been played, but we will still be looking at a book the size of a telephone directory. As an aid to playing the game—telling us what move to make next to have the best chance of winning, even against a randomly playing god—it will be pretty much useless, at least without the aid of a computer.

Fortunately, patterns in the data soon become apparent. For example, not all the words appear equally potent. Whichever god says ag often wins that game. Flump, choo, nis, and doh are not nearly so useful; wibble, fizz, gah, and zig come somewhere in between. Eventually you work out the secret: there are eight "magic triples" of syllables, namely,

xig, flump, wibble
ni, ag, choo
gah, doh, fizz
xig, ni, gah
flump, ag, doh
wibble, choo, fizz
xig, ag, fizz
wibble ag, gah

The first god to include a complete magic triple in the words it has spoken—it does not matter in what order the words of the triple are called, or whether it says other words in between them—wins. This is in a sense a complete description of the game, and it is certainly more compact than the telephone-directory-length list of all possible games. Moreover, because the list of winning strings exhibits certain symmetries, the expedition's mathematician could find ways to code the information still more compactly. But you still have no feel for what is going on in the gods' heads as they play, and little confidence that you will win when the time comes for you to leave the ship and challenge one of them yourself.

Then the expedition's physicist comes to you. "I have it!" he shouts triumphantly. "The aliens are playing a simple game with sticks. Here is my interpretation.

"They start with an imaginary pile of nine sticks measuring from 1 unit to 9 units in length. Each alien claims a stick from the pile by calling its length. I have cracked the code for the alien number system." He writes down the following table:

flump = 1
gah = 2
choo = 3
fizz = 4
ag = 5
xig = 6
ni = 7
wibble = 8
doh = 9

"Each alien takes a stick in turn, adding it to his personal collection, until he has a set from which three sticks add up to exactly 15 units in length. Let me remind you of the first game we saw, the game that went: XIG flump WIBBLE ni AG choo DOH gah. After those moves, the first alien has chosen sticks of length 6, 8, 5, and 9 units. No three of these add to 15. The second alien has sticks of length 1, 7, 3, and 2 units. No three of these add to 15, either. But then the first alien says FIZZ and claims the 4-stick. Now 6 plus 5 plus 4 equals 15, and he duly wins. I have solved the mystery: the aliens are playing the one-dimensional game Fill-the-Gap."

You are in the middle of congratulating him when the cabin boy bursts in.

"Captain, I have solved it," he shouts. "The aliens are playing a simple two-dimensional game! I have cracked the code for the alien position system." He shows you the following table:

xig	flump	wibble
ni	ag	choo
gah	doh	fizz

"Really, sir, all the aliens are doing is playing tick-tack-toe. Remember the first game we saw, the game that went XIG flump WIBBLE ni AG choo DOH gah? After those moves the board looks like this, where the first alien writes X and the second O:

X	O	X
O	X	O
O	X	

"So far, neither alien has a line of three. But then the first alien calls 'FIZZ' and wins with a diagonal line."

You scratch your head, completely baffled. Both of them seem to have an equally strong case. To whom are you going to give the bottle of whiskey you have promised as a prize? Are the aliens really playing a one-dimensional or a two-dimensional game?

As you ponder the matter, there comes a knock at the door: It is the expedition's archaeologist.

"I think I can help," he says. "I remembered the famous Rosetta stone, which carried the same message in three languages. It inspired me to draw a tick-tack-toe board labeled as follows."

6 xig	1 flump	8 wibble
7 ni	5 ag	3 choo
2 gah	9 doh	4 fizz

"Why, of course," you exclaim, "it is the famous magic square: one whose every row, column, and diagonal add to 15. With this board, you can see instantly how the numerical and tick-tack-toe interpretations of the alien game are really one and the same thing. It was obvious that this had to be possible, when you think about it!"

The archaeologist departs rather hurt, but your problem of who wins the prize is not solved. So you call a meeting of the entire crew. The physicist and the cabin boy present their respective interpretations to general applause. But then the expedition's anthropologist stands up.

"I have a better interpretation," he says. "These aliens are gods who would never bother with anything as trivial as tick-tack-toe. What matters to gods is worshippers. These gods are obviously picking sacred triads of worshippers from a species that has three sexes, male, female, and neuter; and three hair colors, black, blonde, and red. They are calling the names of a group of nine people who between them possess every combination of these characteristics.

"It follows from universal aesthetic laws that a god would want a triad of worshippers who are either as alike as possible, or as different as possible. Either three people all of the same sex, but who must have different hair colors; or three people all with the same hair color, but who must be of different sex; or, at the opposite extreme, three people each of different sex and different hair color than the others. A group of the latter type must, however, include ag. I can tell from universal psycholinguistic principles that xig, flump, and wibble are male; ni, ag, and choo female; and gah, do, and fizz neuter. Xig, ni, and gah are blondes; flump, ag, and doh redheads; and wibble, choo, and fizz dark-haired. Ag is uniquely important because her fiery hair and femininity symbolize the importance of contrast. I can also tell from universal aesthetic principles that the gods are imagining their worshippers gathering in a vestibule lined with red velvet, and choo is wearing a bronze amulet."

Before you can comment on this, someone stands up and beats you to it. He is the expedition's cultural relativist.

"You are all talking ze rubbish!" he says in a French accent. "You cannot possibly know what is going on in ze minds of zese alien gods. All you are doing eez projecting your own cultural prejudices. Trying to find what game zese aliens are playing is a futile exercise! You might as well choose any story."

You can sympathize with his sentiments when it comes to the anthropologist's absurdly rococo tale with its wealth of unverifiable detail. But surely he is being a bit hard on the other two interpretations? Then it occurs to you that they are, indeed, also colored by cultural subjectivity. The fact that the cabin boy used X's and O's as symbols on the tick-tack-toe board was certainly culturally determined; and you need two kinds of symbols or objects that are easily distinguishable from one another to play tick-tack-toe sensibly, whatever choice you make. Even the adds-to-15 system was culturally influenced. Why choose positive integers from 1 to 9? You could number the squares instead from 0 to 8; then each line would add to 12. Or you could number the squares from −4 to +4, so that each line adds to zero. And why choose consecutive integers at all . . . ? The only way to avoid spurious cultural overtones is to stick to the aridity of the mathematician's minimalist algorithm, and not attempt to visualize.

Certainly, the cultural relativist is mistaken to claim that any story will do, because stories contain statements that can be falsified as well as ones that cannot. The anthropologist might have gone a bit over the top about the red velvet, but the eight winning triads described by his theory are the correct ones. There might be many correct tales to choose from, but there are even more incorrect ones; for example, any tale that predicted xig, choo, and doh were a winning triad would be wrong.

But it still looks as if you will have to split the bottle of whiskey not merely three ways, but potentially infinite ways, because games theory tells you that there is no end to the supply of correct games that can be invented. Then it occurs to you that there is a point to all this. You want to go out there and kick some alien ass, preferably all by yourself without the help of a computer. From that point of view, there is no question which interpretation of those you have seen is the winner. Human beings have evolved to be extremely good at processing two-dimensional patterns—and relatively weak at arithmetic and abstract logic. In this instance, tick-tack-toe should be your choice of arena. It is the cabin boy who should get the whiskey.

In the world of real physics, of course, we are not dealing with a two-player game like tick-tack-toe. Real physics is more like playing solitaire, seeing what cards turn up. But an immensely empowering insight follows from our study of games: There are bound to be many equally valid ways to look at the universe. We are thus free to pick whichever one we find most comfortable and useful to work with.

Note that I have not claimed—although many have—that the current interpretations of quantum theory are as equivalent as the superficially different forms of tick-tack-toe above. The question of whether they are experimentally distinguishable is one that we will address in the later chapters. The point I am making is that whatever further experimental discoveries there may be, we will always have a choice of ways to visualize the universe. It is our duty to our successors, to those who must go out and battle the laws of physics on territory beyond that currently explored, to make that choice the best one we can at every stage.

It is wonderful to know that with sufficient ingenuity, there is almost no limit to the number of stories we can use to describe the universe. But there is a downside to such ingenuity. Its application can also twist a bad story, one that is not a good way to look at things, so as to make it irrefutable. We will look at two great cautionary examples from the history of science. Both are now generally described as disproved theories, but I will argue that they are merely inept interpretations. They cannot be proved wrong and it is only too plausible that if their proponents had been a little more ingenious, they might still be accepted wisdom—and our understanding of physics would be immeasurably poorer.

Cautionary Tale 1: Phlogiston

The first of these old interpretations is the concept of phlogiston, sometimes also referred to as calistogen. Phlogiston was postulated as an invisible substance that permeated all solids—and indeed all liquids and gases. It conferred the property of heat; the more phlogiston an object contained, the hotter it was. To phlogiston, all substances were porous, so that whenever you put a hot object in contact with a cold one, phlogiston flowed naturally from the hot to the cold until both had the same temperature, just like water flowing to equalize its level or gas to equalize its pressure.

Before the age of machinery, phlogiston really worked very well as an explanation of heat. Different substances differed in the amount of phlogiston they could contain per unit volume. In modern terms, we would say that they have different specific heats. Substances also differed in how readily phlogiston could flow through them; in modern terms, they have different thermal conductivities. That was natural enough; different substances also have different capacities to absorb and permit the flow of ordinary liquids through them (contrast what happens when a sponge, a book, and a house brick are placed in a bucket of water). Phlogiston was compressible, but it possessed some kind of volume, because substances expand when they are heated. Indestructible phlogiston explains why heat is conserved—and to early scientists, it did seem to be conserved, because devices for turning heat into mechanical work functioned at extremely low efficiencies.

One problem with phlogiston was that it did not appear to have any detectable weight. But a far more serious difficulty became apparent with the start of the industrial age—and that was the apparent ability of machines to create new phlogiston. A turning shaft can generate unlimited heat at a point by means of friction, and this works even if the shaft is made of an insulating material so that little or no heat can flow along it. This simple fact was the downfall of the concept of phlogiston.

How lucky that its defenders were not as clever as modern philosophers of physics. If they were, they could have easily explained the apparent problem away. Because, of course, the shaft must have some device at the other end to turn it; for example, a steam turbine takes steam in at a high temperature and ejects it at a lower one. At that end of the device, heat is consumed and phlogiston is apparently vanishing. The process could be explained by the hypothesis of phlogiston tunneling—assuming that phlogiston just undetectably and instantly jumps from one place to another. Does this remind you of something?

Nowadays, we can even in a sense verify that phlogiston has weight. Einstein's famous $E = mc^2$ equation predicts that energy has mass, and this includes heat energy. If you take two otherwise identical objects, each containing exactly the same number of atoms, the hot object does in fact weigh slightly more than the cold one. The difference was simply too small for 19th-century instruments to measure. We arguably had a very near miss with getting stuck with the notion of phlogiston, and failing to progress to the more general concept of energy.

Cautionary Tale 2: Epicycles

A second famous example of a flawed scientific paradigm is the notion of epicycles. Ancient astronomers, trying to figure out the motion of the planets in the heavens, were handicapped by not one, but two, false assumptions. The first was that Earth was itself stationary at the center of the motion. The second was that the planets, being perfect heavenly objects, must move in circles. Because the planets obviously did not move in simple circles, their motion was described in terms of epicycles. For each planet, an invisible pivot point did move in a perfect circle, and the planet moved (on an invisible arm) about the pivot point in a second smaller circle. This idea could crudely approximate the motions of the actual planets as seen in the sky, but for greater accuracy it was necessary to postulate second, third, and even fourth epicycles.

Then came Copernicus, Galileo, and Kepler. As everyone knows, the new hypothesis, that the Sun was the true center of the system, with the other planets including Earth orbiting around it, displaced the old assumption. What is not so well appreciated is that the idea of epicycles need not have died at that point. Kepler discovered that the planets do not orbit the Sun in perfect circles, but in ellipses; and they do not move at constant speed, but faster when they are nearer the Sun, and slower when they are farther away. But this kind of motion can be explained quite well in terms of epicycles. If we assume that each planet has an epicycle whose diameter is equal to the difference between the planet's nearest and farthest distances from the Sun, and that the direction of the epicycle is retrograde—that is, it turns in the direction opposite to the motion of the main arm—then the planet will indeed move fastest at its closest approach to the Sun, and more slowly as the distance increases. We are lucky that Kepler was a stickler for accuracy, and rejected this tempting fudge.

How fortunate it is that he did not have access to modern mathematical techniques and computers. Because we now know that just as a technique called Fourier analysis can approximate a two-dimensional graph as a sum of an infinite series of sine waves (a technique often used in applied mathematics and engineering), so any three-dimensional motion—not just elliptical orbits—can be approximated to any desired degree of accuracy as the sum of an infinite series of circular motions. A sufficiently clever mathematician could even work out a formula for predicting the epicycles of an object, like a comet or spacecraft, entering our solar system for the first time. If the mathematics of Kepler's day had been advanced enough, we might have been stuck with the concept of epicycles.

This would have produced an odd puzzle when relativity was discovered, because in some circumstances (planets orbiting a neutron star, for example), the imaginary pivot points of the epicycles could perfectly well be moving faster than light. Scientists would struggle to explain how, although the invisible arms propelling them moved faster than light, the motions of the epicycles fortunately always seemed to cancel at the right moments, so that the actual planets never broke the speed limit.

Do epicycles remind you at all of the imaginary waves of quantum?

But enough of the negative. Before we pick our favorite story of quantum, let us look at the approaches that have worked well in developing the other aspects of the scientific world-picture we accept today.

CHAPTER 8

A DESIRABLE LOCALITY

We have a choice of stories to tell ourselves about quantum, a choice of arenas in which to play physics against the gods. But what gives us an expectation that a straightforward account is possible? Surely physicists, of all people, have of necessity long been accustomed to accepting esoteric and unlikely stories?

Well, actually, no. For at least 2,000 years, right up until quantum came along, science had progressed by taking exactly the opposite attitude—that the universe should be understandable, and that we could find straightforward ways to visualize what is going on. Nay-sayers—those philosophers who pointed out, rightly enough, that there is no reason the universe needs be comprehensible even in principle, let alone by our limited minds—were cheerfully ignored.

And the approach worked spectacularly well. Blindly optimistic though it was, the expectation that the universe should conform to simple principles that were not only understandable, but even aesthetically pleasing to our ape-evolved brains, yielded breakthrough after breakthrough. So the frustration many physicists now feel about being unable to understand quantum is not the mild disappointment of a gambler whose ticket has failed to win the lottery. It is the fierce rage of a player who sees his winning numbers come up one after another—then gets home only to discover that Schrödinger's cat ate his lottery ticket.

However, the statement that, without quantum, the rest of modern physics is easy to accept needs a little justifying. There is a myth about scientific progress that goes something like this:

In the good old days—say, around Isaac Newton's time—the laws of physics conformed to reasonable intuition. All objects, from billiard balls to planets, moved and interacted in a logical fashion in a universe that was easy to visualize. But then special relativity was invented. We had to accept that basic intuitions about space and time hard-wired into our brains were wrong. General relativity made matters worse still. When quantum theory joined the trio of new understandings, it merely underscored the lesson: The universe can be understood only in terms of highly abstract concepts. Let's face it, three strikes and we're out—we'll never get back to a simple world-picture we can visualize. It can only get worse from here on.

This myth is totally misleading. In some very important ways, the development of special and general relativity actually *restored* a simple intuitive picture that had been wobbling ever since Newton. And that leaves quantum sticking out like a sore thumb.

However, there are several aspects of modern physics that are admittedly a little startling at first encounter. So before trying for an intuitive picture of the universe that includes its quantum aspects, let us first perform a limbering-up exercise. If we overlook quantum weirdness, can we visualize the world without difficulty, including its relativistic aspects? In what follows, please keep a careful watch for the following distinction: Is the world behaving weirdly? Or does it just *look* as if it is behaving weirdly, as we see it from unaccustomed perspectives?

Hello, World

As we go from newborn babe to adult, our worldview gets refined by successive approximations. Later, it is easy to forget how hard the early stages were, so we will take things right from the start.

We are born with the laws of physics already programmed into our brains, at least after a fashion. We know this because of an ingenious technique called the gaze test, invented by developmental psychologist Karen Wynn. You cannot ask a day-old baby what it is thinking. But babies, almost from the moment of birth, gaze in curiosity at the world about them. If something happens in front of them, they normally watch as events unfold, then look away. However, if something occurs that the baby finds surprising, it stares for a much longer time. These "gaze time" measurements are quite objective and can be recorded on videotape for later checking, so these data on baby thought processes are much more reliable than investigations that rely on anecdote, or on the mother's interpretation of early-stage baby talk.

For example, suppose an experimenter places three apples on a tray, then lowers a curtain that blocks the baby's view of it. The (empty-handed) experimenter approaches the tray and fiddles about with the contents, then withdraws, and raises the curtain again. If there are still three apples on the tray, albeit in different positions, the baby glances briefly at them, and then its attention wanders to other things. But if there are now two apples, or four, the baby stares. And stares. And stares.

Similar simple conjuring tricks establish that babies have a whole set of built-in expectations about the world. For example, they differentiate between the animate and the inanimate. Using criteria that are not yet wholly clear, they place the things they see into either the class of the animate (objects that have the power to move themselves and things they come into contact with), or the class of the inanimate (objects that are inert). Thus, a baby is mildly interested but not astonished when a sleeping cat wakes up and walks away, or when a human pushes a building block across a table with her finger. But if the building block starts moving apparently of its own accord, the baby gazes in wonder.

These tests prove more than is at first sight apparent. For example, the first test is used to establish that babies have the innate ability to distinguish numbers up to about four. But it also demonstrates that babies start with a built-in expectation of conservation laws—apples, or other objects, do not simply pop in and out of existence. Nor can they be teleported; otherwise, any missing or extra apples could simply have been transmitted to or from somewhere out of the baby's sight. An object is expected to move only by means of a continuous progression. (I am tempted to say that even a day-old baby knows that the science of *Star Trek* is nonsense.) Likewise, the test with the building block proves not just that a baby distinguishes between animate and inanimate, but more subtly that it expects objects to affect one another only when they are in physical contact. A baby is not surprised when a block moves if somebody's finger is touching it.

A key concept is already emerging here: *locality*. Objects move locally, rather than jumping around in space, and they interact locally. Indeed, if it were otherwise, it is difficult to see how a baby could make any progress in comprehending the physical world. Of course a baby's brain does not have all the information it needs about the world preprogrammed into it—far from it. The built-in expectations serve as a kind of bootstrap, an outline framework of rules that will be repeatedly refined and modified. For example, a baby has a built-in expectation that objects, including itself, will fall unless they are supported by other objects. Yet in due course, it learns to accept birds, balloons, and aircraft as exceptions to the rule. This progress, modifying our ideas as we go along, continues for quite some time. As we grow up, the data we get from personal experimentation, such as pushing our toys about, are increasingly supplemented by information taught to us by others. The next section roughly charts the stages by which the worldview of a modern child progresses. Just as the development of the human embryo approximately recapitulates our evolutionary history—for example, at one stage it has gills—so the child's conceptual progress approximately reprises the historical stages by which scientific understanding has progressed.

Worldviews, Infant to Adult

Nursery Physics

The world is a flat and stationary surface that goes on forever. The Sun, Moon, and stars are high up above, stuck on some kind of invisible dome. An invisible force pulls everything in the same downward direction, making any object fall unless it is supported by something else. Inanimate objects stop moving as soon as you stop pushing them. Objects have definite positions. Objects can affect one another only if they are touching.

Elementary School Physics

The world is round. Gravity pulls you toward the center wherever you are. The Earth, Sun, Moon, and countless stars all hang floating in a space of three dimensions, with the Earth turning as it goes round the Sun. Objects keep moving in the same direction unless some force, such as friction, stops them. Objects have definite positions. Time is marked by clocks, and events happen at definite times. Objects can sometimes affect one another without touching, for example, by electrostatic forces, by magnetism, or by light or radio waves—these are all encountered as different phenomena.

High School Physics

As well as physical matter, space contains invisible spread-out entities called fields. An electric charge creates an electric field around itself. Moving electric charges (such as the current in a wire) create magnetic fields. Accelerating electric charges (such as the alternating current in a radio antenna) generate waves made up of rapidly varying electric and magnetic fields that travel at the speed of light. Electricity, magnetism, and electromagnetic waves are merely different aspects of the same phenomenon.

College Physics

There is no such thing as absolute rest. Measures of distance, such as the distance between two stars, and times, such as the time lapse between two events, depend on the motion of the observer. Time can pass at different relative rates for different observers. The structure of space-time is warped by gravitation. Any object falling under gravity, such as a planet orbiting a star, is actually traveling in a straight line relative to the region of space immediately surrounding it. Nevertheless, any one observer still sees that all objects occupy definite positions, and all events happen at definite times.

Scary Physics

The world is described by the equations of quantum mechanics, which don't tell you for certain what is where. Objects can no longer be thought of as having definite positions or speeds. Maybe even cats can no longer be thought of as definitely alive or dead.

Join the Flat Earth Society!

We have all had to revise our ideas many times in order to attain a proper grasp of physics. Each time, some things that were previously believed true had to be accepted as false, or at most as mere approximations to the new, better truth. Luckily, human beings seem to cope remarkably well with learning in this way. Serious students of almost any subject, not just physics, become hardened to hearing a lecturer say: "Everything you were taught last year was nonsense, a story designed to prepare your mind for the real truth, which is as follows. . . ."

However, not everyone can ascend the paradigm ladder successfully. Back when I was a college freshman, a fellow student jokingly wrote to the agony column of one of Britain's tabloid newspapers along the lines of:

"Dear Marje, I believe the earth is flat, and my friends make fun of me for it. Please help me."

Back came the reply:

"Do not worry. There are many people who feel exactly as you do. . . ." The letter went on to give details of a Flat Earth Society, which then met weekly or monthly in London. This was some 20 years ago—already quite some years after beautiful, high-quality photographs of the round Earth taken by various sets of Apollo astronauts had started appearing in practically every newspaper and magazine on the planet. I recently went looking for Flat Earth societies still in existence with a view to interviewing some of their members. Alas, either the prevalence of orbiting astronauts or other factors appear to have finally put paid to this view as an organized school of thought. Although you will find many spoof references on the Web, the last sincere Flat Earth Society seems to have expired some years ago.[1]

There are useful lessons to be learned from the flat-earth hypothesis, however. Because almost nobody nowadays feels threatened by the concept that the world is a sphere, we can look at the difficulties in going from one worldview to another more clearly and dispassionately than might be the case with some of the steps we will have to take later. The durability of flat-earth belief shows how hard it can be to accept a new concept, even one that is well within our capability to visualize and does not contradict the evidence of our senses. After all, we understand from infancy that the universe is a three-dimensional place, containing three-dimensional objects. It is also basic to understand that if a circle or sphere is very large, its curvature is very small, so that the curvature of the Earth is not easily noticeable by inspecting your immediate neighborhood. Yet it can still be disconcerting to abandon the "world is flat" view, which starts as our default perspective.

I happen to be able to remember unusually far back into my own childhood. I know this because when I was two, my family traveled to Australia and back by ocean liner, and I have clear memories of the voyage, the only one we took during my childhood. I can remember how profoundly disconcerted I was to be told that, even though people in Australia were standing upside down relative to people in England, they did not fall off the world because "gravity is a force like magnetism that pulls you toward the middle of the Earth wherever you are." I thought that when we arrived in Australia I would *feel* upside down, but much like a character using magnetic boots to walk on the ceiling in a cartoon film, there would be a spooky force pulling my feet up toward the ground. Told that the Earth was turning and rushing through space at great speed, I went down to the bottom of our garden, far from the noise and vibration of the road traffic. Even there, I could not feel the slightest sense of motion.

By now, you are probably smiling at the naivete of my two-year-old self. You no doubt have a clear mental image of the Earth as a sphere, illuminated by sunlight on only one side at any given moment, and pulling everything on its surface toward its geometrical center by the force we call gravity. Of course, the local direction of downward, and the local time of day, is different on each part of the surface.

Some readers will remember their own discomfort and amazement at first being told that the world was round. It is an historical fact that the notion of the flat Earth clung in many people's minds thousands of years after scientists knew that it was spherical. Historian Jeffrey Russell has thoroughly debunked the myth that the ancient Greeks' discovery that the Earth was round was either forgotten, or opposed by any mainstream church, during the so-called Dark Ages.[2] Throughout recorded European history, mainstream natural philosophers have never seriously doubted that the Earth is a sphere. The only informed debate in Columbus's day concerned exactly what the diameter was. But that did not stop huge numbers of people preferring the notion of a flat Earth, almost up until the present day. Revising one's ideas can be painful even when the new picture is well within our intuitive capacity to grasp.

Action at a Distance

In historical terms, the junior school period represents a giant leap forward: from the Middle Ages to the Newtonian worldview, which dominated from the 17th to the 19th centuries. I suspect that to many people, this intermediate period represents a kind of golden age or comfort zone. The workings of the solar system, the nature of gravity, the basic rules of mechanics involving momentum and friction, were very well understood. But no one stopped to worry overmuch about the nature of space and time, which were assumed measurable with respect to some kind of absolute grid or framework. We lived in three fixed dimensions of space, and one of time, and that was that. And although some questions about the nature of light, and a few oddities like magnetism, remained obscure, these were mere details that could be overlooked.

I would beg to disagree. The deceptively friendly Newtonian picture actually robs us of something beyond price, the key feature whose assumption enabled us to make sense of the universe from the cradle, and that is what I shall call *the principle of locality*. That is a more formal name for the rule, "Things affect other things only when they are very close to them." In our cradle physics, the force of gravity was not troubling from that point of view, because it was assumed to be universal and unidirectional, not coming from any particular object. But once the force of gravity is understood as due to the Earth's mass— in fact, the sum of tiny attractions from all the trillions of quintillions of quintillions of particles that the Earth comprises, acting over a range of many thousands of kilometers—then we have the phenomenon that Hooke and Newton called "action at a distance." Newton assumed correctly that gravity had an effectively infinite range, becoming weaker at great distances, but never reducing to zero, but he also assumed incorrectly that its effect was instantaneous—so that moving an object a million miles away would instantly change the effect its gravity had on Earth.

The Friendly Field

Gravity was not the only nonlocal force in the Newtonian world-picture. Two other kinds of action at a distance were also known, although they appeared to affect only certain kinds of matter. These were the forces that we nowadays call electrostatic and magnetic. Although both phenomena had been studied before, magnetism by Gilbert and electrostatics by Charles Du Fay, British scientist Michael Faraday's research in the early 19th century went much deeper.

School pupils today still learn of Faraday the experimenter, investigating the intimate relationship between electricity and magnetism, but it is less well known that his deeper motivation was philosophical. He was profoundly and instinctively opposed to the notion of action at a distance, and wondered if electric and magnetic forces could be explained in any other way. This led him to the concept of lines and fields of force. As a simple example, consider the two pith balls shown in Figure 8-1.

Both balls in Figure 8-1 are positively charged and they repel one another as shown. One way to think of what is going on is that each acts on the other directly at a distance, as indicated by the double-

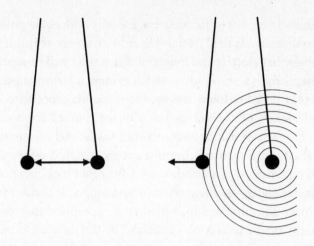

FIGURE 8-1 Do charged objects repel one another by direct force, as on the left, or via fields, as on the right?

arrow on the left. But an alternative interpretation, shown on the right, lets you consider that each charged ball surrounds itself with an invisible field that extends through space, which we would nowadays call an electric field. Each ball is pushed, not directly by the other, but by the field that surrounds it. Similarly, you can think of magnets either as operating on one another directly, or as being surrounded by magnetic fields.

Faraday's contemporaries were initially scornful of his field notion. It seemed to violate Occam's razor; why postulate an unnecessary, invisible entity? Some modern philosophers of physics might have dismissed the idea for a rather different reason, that the question of whether the electric field was real or not was merely a matter of *interpretation.* If an electric field is discernible only by the force it exerts on a charged body, then surely the question of whether the field is "really" there when no charged body is present is an untestable, angels-on-the-head-of-a-pin kind of proposition. We are therefore free to think of electric forces in terms of action at a distance or in terms of fields, as we please. The only thing that deserves to be called real is the mathematical algorithm that enables us to calculate the forces exerted, the inverse-square rule.

Fortunately for scientific progress, Faraday was not sophisticated enough to dismiss his field notion for either of these reasons. He felt very strongly that electric and magnetic fields were real things. And as so often happens in science, what started as mere interpretation turned out to have real and testable consequences. Faraday speculated that if a field had a reality of its own, then moving the source need not change the state of the whole field instantaneously. Just as real substances have finite elasticity, and transmit impulses at finite speed—for example, when you tap one end of a wooden ruler, the other end does not move until an instant later—so might electric and magnetic fields. Faraday's extraordinary intuition led him further, to speculate that radiation such as light might in fact be vibrations in the lines of force of his field, that gravity also might be transmitted at finite speed through the medium of a field, and even that the particles of which matter is made might be no more than knots in these fields.[3] Arguably, he thus predicted important elements of both special and general relativity, and even string theory.

However, Faraday lacked the mathematics to develop his predictions quantitatively. This was done by the Scottish scientist James Clerk Maxwell. Like Einstein, Maxwell was primarily a visual thinker. His insights were developed in terms of lines and areas, surfaces and volumes, topology and geometry. Although he was very competent in math, it was his servant, not his master. The entities that he described had to have visualizable meanings, even though they described invisible things—a lesson for today's quantum physicists. Thus he was soon deriving such useful quantities as magnetic pressure, measured like ordinary pressure in pounds per square inch, and magnetic energy. It turned out that a strong magnetic field could be thought of as storing energy, just like a compressed gas, so much per unit of volume.

There is a symmetry between electric and magnetic fields that is normally obscured because in the laboratory we can find plenty of particles carrying an electric charge—protons and electrons—but no corresponding ones with magnetic charge. Nevertheless, an electric field can also be created by a change in a magnetic field, and vice versa, in a yin-and-yang relationship. This led Maxwell to an intriguing possibility: Could you create an electric-magnetic field that existed independently in its own right, with no associated physical object? The answer turned out to be that such a phenomenon could exist, but would never be stationary: It would propagate through empty space like a wave or ripple at a speed that was very high, but could be calculated from two known electrical properties of the vacuum, called the permittivity and the permeability. The speed of the predicted wave exactly matched the measured speed of light.

Maxwell died tragically young, at the age of 48, but his famous "Treatise on Electricity and Magnetism" was developed by colleagues George Fitzgerald, Oliver Heaviside, Oliver Lodge, and others into a complete and beautiful picture. Electric and magnetic fields can be thought of as represented by little arrows having magnitude and direction—we now call them vectors—associated with every point in a volume of space. If you draw an imaginary surface around that volume, then the difference between the quantity of flux arrows going into and out of the surface defines the net amount of electric charge within it. More complicated geometrical calculations yield more subtle quantities, such as the energy associated with a given volume of a magnetic field. And so we can design the electric generators, motors, and many other devices on which our modern civilization depends.

But what is important to us is that, at least as far as electric and magnetic forces were concerned, Faraday and Maxwell had abolished action at a distance and restored locality. They had demonstrated beyond reasonable doubt the existence of fields—invisible entities that were real enough to contain energy of their own—and that objects interacted not with far-off things, but only with the electric and magnetic fields immediately surrounding them. There is no instantaneous electromagnetic interaction at a distance: With sufficiently delicate instruments, you might be able to detect the field due to a magnet a million kilometers away from you, but if somebody suddenly moves that magnet, the magnetic field around you will not instantly change. Any such change can propagate out only like a ripple at finite speed, and the maximum speed is, by definition, the speed of light, the speed of an unencumbered electromagnetic wave in free space.

There are at least three reasons to celebrate Faraday's and Maxwell's abolition of the action at a distance of the Newtonian picture. The first is simply that such remote action is upsetting to our early intuition, our instinctive baby expectation that objects can interact only by touching. I can still remember how spooky it was when, as an infant, I was first shown how a horseshoe magnet could snatch up its keeper bar when it was still a good inch away from it. The touching rule is of course only a rule of thumb that evolution has found advantageous to install in us, but it is a very useful rule for simplifying what might otherwise be an incomprehensible world.

The second reason is the theoretical worry that if objects can directly affect one another from far away, it undermines the hope that we can ever properly test the laws of physics by experiment. If the actions of processes on, say, Alpha Centauri can directly and instantly affect the behavior of equipment in a terrestrial laboratory, raising the possibility of self-amplifying feedback interactions, then we can never perform an experiment on a truly isolated system.[4]

The third reason is simply that instant long-range interactions make it much harder to construct simple predictive models of the world. This applies to all kinds of models, including traditional ones based on fearsome-looking differential equations, but is easiest to see by using a more modern device, the cellular automaton. Ever since the computer was invented, the cellular automaton has been the physicist's tool of choice for modeling systems that occupy an extended volume of space—which is to say, just about everything the real world contains, be it a solid, liquid, gas, or something more exotic. A simple example is shown in Figure 8-2, which depicts fluid flow.

We can calculate forward from the picture on the left to that on the right quite economically, provided that each cell is directly influenced only by the cells immediately around it. For example, to find the new state of the cell which is shown shaded on the right, we need take into account the previous state only of the cell itself and its immediate neighbors, as shown lightly shaded on the left. If nonlocal influences were at work, we would have to take into account the state of all the other cells, in principle extending an indefinite distance in every direction, and the amount of calculation involved would be vast. If on the other hand there are no nonlocal influences, it opens the door not only to the idea that the universe can be economically modeled, but

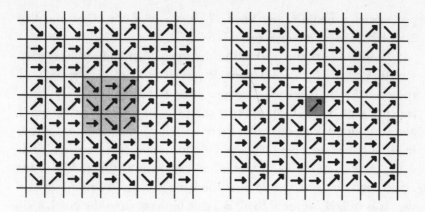

FIGURE 8-2 Modeling the flow of a turbulent fluid by computer simulation: one time step takes you from the left picture to the right one.

even to the possibility that it might actually *be* something as understandable as a kind of local cellular automaton, a hypothesis we will return to in the last chapter.

Faraday himself will remain an inspiration to us in two ways. First, the specific concept of locality, that forces can operate only on nearby things, has turned out to be of immense importance. But even more, his attitude—a stubborn practical man's insistence that the universe shall be intelligible, and shall conform to our notions of common-sense, however difficult this goal might sometimes seem—will guide us in our quest.

But now it is time to graduate from high school. . . .

A Moving Perspective

Maxwell's brilliant work had of course left one interpretational question hanging: Given that light is an electromagnetic wave, in what medium can the wave be considered to be traveling? After all, sound waves are a movement of air molecules, and sea waves a movement of water particles; even though light waves are rather more abstract, surely they must travel in some kind of supporting medium? As far back as the mid-18th century, the great mathematician Euler had hypothesized a medium that filled all space, called the ether, and that "sunlight is to ether what sound is to air."

However weak the interactions between ether and ordinary matter, ether should have at least one detectable property—its speed with respect to Earth. To see why this is so, imagine that the ordinary atmosphere has no detectable properties except for its ability to carry sound, and you want to know whether there is a wind blowing. Sound travels in air at 330 meters per second, so if you station observers in a circle 3.3 kilometers in radius and set off an explosion in the middle of the circle, each observer should hear the bang exactly 10 seconds later if the air is still. However, if there is a gale blowing from the north at 30 meters per second, the sound reaching the northernmost observer in the circle is delayed by about 1 second, taking 11 seconds to reach him, whereas it will reach the southernmost observer 1 second early, after only 9 seconds. In fact, a wind of any speed and direction causes some observers on the circle to hear the sound earlier than others.

In exactly the same way, any ether wind with respect to Earth's surface should be detectable because light would travel slightly faster in some directions than others. No one knew whether the solar system was moving or stationary with respect to the ether, but because the Earth orbits the Sun at some 30 kilometers per second, continually changing direction as it does so, it could not possibly be stationary with respect to the ether the whole time. A variation in the apparent speed of light on the order of 1 part in 10,000 should have been easily detectable with late Victorian instruments.

It was Maxwell who first described a practical experiment to detect this ether wind, but he died of cancer before it could be carried out. It is extraordinary to think that had he lived a little longer, he might well have anticipated Einstein in the development of special relativity.

Of course, no ether wind could be detected when the experiment was eventually performed by Michelson and Morley. Precise astronomical observations ruled out other possibilities, such as the idea that Earth somehow dragged the local envelope of ether along with it. In that case, the effects of ether current should show up as subtle variations in the timing of such events as eclipses. It could hardly be the case that all the ether in the solar system was being dragged along in step with our particular planet. Similarly, measurements on stars that orbited one another rapidly ruled out the idea that light traveled with a fixed speed relative to its source, like a bullet fired from a gun. There really did appear to be a deep paradox here.

It was of course Einstein who solved it, with the bold postulate that space and time are not absolute, but vary with the motion of the observer in such a way that light always appears to move at constant speed. For example, if a spacecraft were to pass Earth at very high speed, then from our point of view its clocks would appear to be running slightly slow, and the ship and everything aboard it would appear contracted in its direction of motion. Conversely, observers on the spacecraft would perceive the rest of the universe as spatially distorted relative to our viewpoint. In general, we would not agree with the occupants of the craft on either the distances and directions of objects or the timings of events that we could both observe.

These effects sound very bizarre, but the apparent distortion of objects moving at very high speeds is really just an unfamiliar kind of perspective. Even the most basic rule of perspective—that faraway objects look smaller—is not hard-wired into our brains. Here is a true account, from an anthropology textbook, of a Bushman who was brought outside his native forest for the first time in his life.

> Turnbull studied the Bambuti pygmies who live in the dense rain forests of the Congo, a closed-in world without vast open spaces. Turnbull brought a pygmy out to a vast plain where a herd of buffalo was grazing in the distance. The pygmy said he had never seen one of these insects before; when told they were buffalo, he was offended and Turnbull was accused of insulting his intelligence. Turnbull drove the jeep toward the buffalo; the pygmy's eyes widened in amazement as he saw the insects 'grow' into buffalo before him. He concluded that witchcraft was being used to deceive him. [5]

Does special relativity make reality harder to visualize? I would argue that it does not in any fundamental way, because we already had to get used to the fact that objects look different from different perspectives, and that different observers might naturally have used different coordinate systems, long before relativity came along. Special relativity asks us to take only one small further step—to the idea that the observer's natural coordinate system and perspective viewpoint vary not just with position but also with speed. The universe can look different even to two people in the same place, if they are moving at different velocities—just as we already know that it might look different to observers in different places. But things only *look* different—cause and effect, the flow of events, are the same to all observers. If an occupant of our imaginary spacecraft makes himself a cup of tea, in our telescope we see him putting the kettle on, getting a teabag, and so forth. If he seems to move rather slowly, and the kettle looks rather squashed, this is just an extension to the rules of perspective we have always accepted. From the astronaut's point of view, the kettle is its usual shape and he is doing everything at normal speed.

The aspect of special relativity that initially seems hardest to accept is the idea that time can appear to flow more slowly in a frame moving fast with respect to yourself. You might find it helpful here to consider that Doppler effects would produce similar oddities even in a nonrelativistic universe. First, consider sound. Suppose that a train is traveling toward you at one-tenth the speed of sound, 70 miles an hour. You will hear the pitch of its whistle as about one-tenth higher than it really is. If your ears were good enough to hear a conversation taking place aboard the train, the pitch of everybody's voice would also sound higher, and moreover, you would hear 10 seconds' worth of conversation in only 9 seconds, because the sound of the last word would have less distance to travel to reach you than the sound of the first word, and so would reach you in less time. If you were blindfolded, it would seem exactly as if life aboard the train were happening 10 percent faster than normal. After the train passed you and was receding, everything you heard would seem correspondingly slowed by the same factor.

In exactly the same way, even in a universe in which light really was an ether wave, life aboard a spaceship coming toward you at a tenth the speed of light would look as if it were happening 10 percent faster than normal—or 10 percent slower if the spacecraft was receding. In our universe, you have to add on the relativistic correction as an additional factor to this Doppler effect, an additional slight slowing of events on board the craft. Actually, even at a tenth the speed of light, the Doppler effect is much bigger than the relativistic correction: It is only at more than half the speed of light that the relativistic correction overtakes the Doppler one. There is really nothing surprising about events in a fast-moving frame of reference seeming to happen at a different rate.

General relativity asks us to stretch our minds a little further and accept that the fabric of space is warped by the force we call gravity, just as the Earth's surface is not flat under our feet but bends slightly. An object falling freely under gravity is actually traveling in a straight line in the warped space that immediately surrounds it.

The subtlety that I think confuses many people, and that is not adequately explained in some of the texts I have seen, is that the warping an object encounters once again depends on its speed as well as its position. For example, consider three spacecraft at a point 100 miles above the Earth traveling at different speeds but all falling freely under gravity. Each travels in a straight line from its own point of view, but with different results, as shown in Figures 8-3a-d. The sounding rocket falls back to intercept the Earth's surface (shown as a thick black line), the orbiting satellite maintains a constant distance from the surface, the interplanetary spacecraft traveling at escape speed increases its distance from the surface. Note that because photons themselves travel so fast compared to Earth's escape speed, what any observer in the vicinity of Earth actually *sees* through a telescope corresponds almost exactly to the "flat-space" view, irrespective of the observer's own velocity. If we lived in the vicinity of a dense massive object like a neutron star, general-relativity-related perspective effects would be familiar to us.

Once we have accepted the new perspective rules, the relativistic universe actually gives us a priceless benefit. It restores locality with an emphasis that has been lacking in every picture since our original nursery physics. Objects interact only with things that they are physically touching—granted that those things are fields rather than physical objects. All forces, electromagnetic and other, exert their effects through the medium of fields, and disturbances in fields—even the rather special distortion-of-space field that is gravity—can propagate no faster than light. There is no action at a distance. And that makes

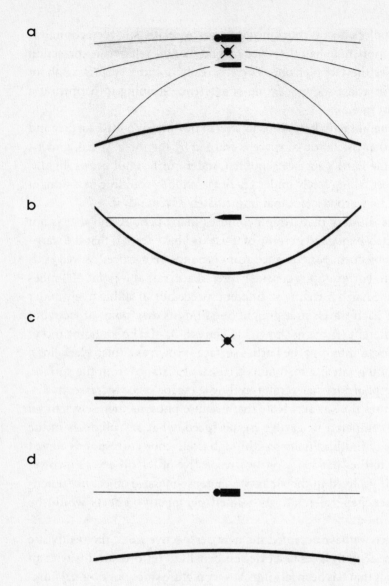

FIGURE 8-3 View of a sounding rocket, a satellite, and an interplanetary space-craft 100 miles above Earth's surface. In fact, each object is traveling in a straight line with respect to its own perception of space.
(a) Flat-space view
(b) Space as experienced by the sounding rocket
(c) Space as experienced by the satellite
(d) Space as experienced by the interplanetary spacecraft

the universe relatively straightforward to understand and model. It is a cosy place in which only things in your immediate neighborhood affect you.

If only we could integrate quantum into this neat local picture, we could perhaps play against the gods on fair terms, in an arena which our brains are wired to understand.

CHAPTER 9

INTRODUCING MANY-WORLDS

Whenever we test a small piece of our universe experimentally, we find that up until that moment it has been behaving as a chunk of Hilbert space, developing not as a single history, but as a nest of interacting probability waves. This description of nature is by far the most accurate that has ever been achieved. Quantum theory makes possible immensely precise predictions of timings and frequencies of microscopic processes, which have been confirmed to an astonishing number of decimal places. The waves of Hilbert space are simply the waves Schrödinger derived a lifetime ago, although we now have better mathematical tools (and of course, computers) to help us work out their behavior. The essence of the many-worlds interpretation is breathtakingly simple. Let us assume that what the math tells us is correct. We can then explain what is going on in terms of three emergent phenomena: *entanglement*, *decoherence* and *consistent histories*.

The math implies that an isolated system, say, a bunch of atoms bouncing around in a sealed container, explores all the ways the atoms might go. It is not a question of atoms surfing guide waves, as in the original picture we tried to construct, but of the atoms themselves trying out all the possible paths, going every-which direction. The waves of Hilbert space explore every possible way the system might develop. If we put a second isolated system containing some kind of observation apparatus in contact with the first system, neither undergoes any kind of collapse. Rather, the mathematics of *entanglement* tell us that the high-probability areas of the joint Hilbert space thus created will develop as *consistent histories*. For example, if the small system is a radioactive atom (one that can spontaneously split apart), and the large system is a radiation detector with an electronic memory consisting of a capacitor that becomes charged when a radioactive particle is detected, then the states atom-split-capacitor-charged and atom-whole-capacitor-uncharged quickly become much more likely than atom-split-capacitor-uncharged and atom-whole-capacitor-charged. Likewise, if the small system is Schrödinger's cat apparatus and the large system is a cat-loving astronaut, live-cat-happy-astronaut and dead-cat-sad-astronaut become much more likely than live-cat-sad-astronaut and dead-cat-happy-astronaut.

Why are the happy and sad versions of the astronaut not aware of one another? The mathematics of *decoherence* tell us that the interference between developing outcomes that are significantly different above the microscopic level fade very rapidly. History lines whose only difference is that one electron has gone through the left slit of a two-slit experiment instead of the right one interfere with one another quite significantly, but history lines where lots of particles are all in different positions (such as the atoms of the cat's body and the electrons within the astronaut's brain in the above example) interfere with one another only to a very tiny extent. As the philosopher Daniel Dennett and others have pointed out, the things that we consider to be real, including ourselves, are simply stable, persistent patterns: The happy-astronaut-live-cat pattern is one such. As far as she—that is, that particular pattern of her—is concerned, she inhabits a single history in which the cat was lucky and lived.

The many-worlds interpretation is sometimes claimed to beat all others by Occam's razor, on the grounds that it requires no new physical assumptions. Accepting it requires only the moral courage necessary to accept that the same rules that apply to small isolated systems, like bunches of atoms, also apply to larger isolated systems without limit, therefore including the largest possible one—our universe taken as a whole.

The no-assumptions claim can be challenged. In fact, even the most prominent supporters of many-worlds nowadays acknowledge that some postulates must be made to accommodate the theory, an issue we'll look at in more detail later. But many-worlds has more going for it than Occam's razor. Chapter 7 prepared us for the fact that there might be many ways to look at physical reality, none uniquely more correct than the others. But many-worlds is still preferable to other interpretations for the same reason that the cabin boy's tick-tack-toe was a better game than the ideas of the other crew members. It is easier for our minds to grasp. It enables us to keep to the intuitive picture that Faraday, Einstein, and other great physicists have struggled to preserve, a universe of three dimensions of space and one of time, in which nothing is random and locality reigns supreme.

That's quite a claim. Let us first lay out the evidence in its favor; we will come to some reservations later. We are now in a position to resolve our Principal Puzzles of Quantum. We will take them in reverse order.

PPQ 4

Why does reality appear to you to be the world in a single specific pattern, when the guide waves should be weaving an ever more tangled multiplicity of patterns?

Answer

Your mind in a specific state is a pattern of information—or speaking physically, your brain in a specific state is a pattern of positions of atoms and electrons. The mathematics of decoherence predict that two brain patterns that initially differ by a trivial amount—say, because one particular photon happened to be transmitted rather than reflected when hitting your cornea, thus reaching your retina—very quickly cease to have any significant effect on one another as the difference grows.

(Oxford philosopher Michael Lockwood prefers *many-minds*. His point: the large Hilbert Space within which all physically possible histories unfold contains mind-patterns that have seen and recorded different versions of events. However this viewpoint leads to philosophical complications; so, I shall stick with a physicist's perspective: A mind is just an information package embedded in a world-line.)

PPQ 3

Why does the universe seem to waste such a colossal amount of effort investigating might-have-beens, things that could have happened but didn't?

Answer

It does not waste any effort investigating might-have-beens. The interference patterns that seem to demonstrate that the universe tried out things that didn't happen—how did the universe know whether the bar-code reader would have registered the chicken going through the other slit?—correspond to outcomes that in fact also happened. However, the world patterns in which they happened decohered rapidly from those in which they didn't as soon as the interaction we call measurement occurred. We now understand that taking information about a system, recording the result permanently in a larger outside environment, is actually what causes decoherence. The terms "permanent" and "larger outside environment" might sound like a cheat, but all I mean by them is an environment containing enough particles that a spontaneous reversal of the recording process becomes unlikely, like the dots of ink on a sheet of paper all just happening to leap back into the bottle they came from.

PPQ 2

Spooky quantum links seem to imply either faster-than-light signals or that quantum events are truly random.

Answer

Assuming many-worlds, the laws of physics do not imply any randomness at all. When, for example, a photon hits a polarizer, the result is quite deterministic. It gives rise to two event-patterns in Hilbert space, one in which the photon is transmitted and one in which it is reflected. There will also arise two different patterns corresponding to the present "you," matching each outcome.

PPQ 1

Spooky quantum links seem to imply either faster-than-light signals or that local events do not promptly proceed in an unambiguous way at each end of the link.

Answer

Locality has always been claimed as a benefit of the many-worlds approach, but the point was not proven until quite recently, in a brilliant paper published in 2000 by David Deutsch and Patrick Hayden.[1] Here, however, we will give a nonmathematical picture of how the correlations of EPR can arise from local effects alone.

To explain the process, we will go back to the lottery cards of Chapter 1 and expand on the notion that causing quantum decoherence—here, by scratching a lottery card and observing whether you get a black or a white spot—gives rise not simply to two worlds, but two *sets* of *local* worlds.

Later, we will consider whether these sets should really be considered infinite, but for illustration purposes we shall assume that each time a spot is scratched, it gives rise to exactly 100 versions of local reality in which the spot is white and another 100 versions in which the spot is black. So when you go into your booth to play the lottery game, when you scratch your card you might think of yourself as creating 200 versions of your booth, each floating around in a grey void a little bit like *Dr. Who*'s Tardis in the old BBC television series. Half of these booths contain versions of you holding a card with a white spot, and the other half have versions of you holding a card with a black spot.

Your partner has similarly created 200 versions of her booth. The subtle bit is how the various booths get allocated to different consistent histories. Here is a crude metaphor for what occurs. Imagine that each version of each booth stretches out a ghostly tendril. At the the end of each tendril is a label with information like, "left booth, spot number 3 scratched, revealed color white." Shortly we are going to use the tendrils to pull together 200 complete classical-looking worlds, each containing one booth with you in it and one booth with your partner in it. We can make the correlations between your and your partner's colors anything we like simply by joining up the tendrils in an appropriate way.

For example, if you have both picked the same spot, we pair the 100 versions of you holding a black card with the 100 versions of your partner holding a black card, and the 100 versions of you holding a white card with the 100 versions of your partner holding a white card. The results all match in all the resultant worlds, as they are supposed to.

If you each pick a spot at 90 degrees to your partner's, we pair the 100 versions of you holding a black card with the 100 versions of your partner holding a white card, and the 100 versions of you holding a white card with the 100 versions of your partner holding a black card. The results are opposite colors in all the worlds.

If you and your partner pick spots one place apart—as you will have if you are trying to win the game—we pair just one version of you holding a white card with one version of your partner holding a black card, and just one version of you holding a black card with one version of your partner holding a white card. Then we pair up the remaining 99 versions of you holding a white card with the 99 versions of your partner holding a white card, and the 99 versions of you holding a black card with the 99 versions of your partner holding a black card. Everyone is accounted for, and you have won in just 2 worlds out of the 200, as expected.

Of course this sorting of diverging worlds does not really involve tendrils with labels on them. It is a process whereby each version of the world containing you comes to be potentially more and more affected by one particular version of the world containing your partner, and less and less by the other versions. You could imagine the process as analogous to pulling entangled skeins of wool gently apart into sheets; or even as resembling the biological process of meiosis, in which chromosomes are duplicated and then in due course spliced back together in appropriately matching ways. But the key point is that nothing happens that would require the propagation of faster-than-light influences. The process of quantum collapse—the process of scratching the card, and even your consciously seeing the result—can happen fast. At that point your Tardis-booth already "knows" what kind of partner is appropriate for it to hook up to. But it does *not* need to exchange information with the maybe far-off partner booth at that point. This is the difference between selecting a partner in a video dating booth, and immediately writing down (or even dialing) their telephone number, which is perfectly possible, and having an actual faster-than-light exchange of messages with your partner-to-be, which is not. Many-worlds respects the spirit as well as the letter of special relativity.

෨෨෨෨

With all this going for it, you might expect that the case for many-worlds would be considered cut and dried. From my perspective in Oxford, where so many of the leading supporters of many-worlds (some of whom we'll soon meet) live and work, it sometimes feels that way. And yet many-worlds is not universally accepted in the world-wide scientific community. Max Tegmark, one of the few leading American physicists who actively supports many-worlds, has published the following results of an informal poll he took at a recent international conference on quantum physics.[2]

Copenhagen: 4—Believers in the modern Copenhagen interpretation in the broadest sense, the idea that the unmodified Schrödinger wave equation gives rise to a collapsed single reality when perceived by a conscious observer.

Collapse mechanism yet to be discovered: 4—Believers in the idea that the Schrödinger wave equation must be modified to include some physical collapse mechanism (for example, Roger Penrose's, which we'll meet in Chapter 14) that gives rise to a single-valued reality.

Pilot waves: 2—Believers in some form of Bohm's pilot-waves notion, that a single reality is traced out by particles surfing on guide waves that in a sense explore all the developments that do not really happen.

Many-worlds: 30—Believers in the idea that collapse never happens, and the universe keeps exploring many different outcomes, which should be considered equally real.

That looks pretty convincing so far: a 75 percent vote for many-worlds, with the opposition split. But there is a further figure: 50 (of the total of 90) physicists in the hall were undecided, or at least unable to agree with any of those four broad choices. That is rather appalling. In one sense, many-worlds is becoming the only game in town. The opposition to it is fragmented and dwindling. But looked at another way, it has a long way to go. Only a third of the specialists in the field were willing to stand up and be counted as many-worlds supporters. Let us look at the reasons—some justifiable, others less so—for this situation.

One problem might be, ironically, that many-worlds is one of those scientific theories that was proposed ahead of its time. Back in the 1950s, before most of the current generation of quantum physicists were even born, Hugh Everett III, student of the famous John Wheeler, wrote a Ph.D. thesis outlining his proposal, which in retrospect seems astonishingly obvious: Why assume that quantum collapse occurs at all? Why not simply believe what the equations are telling us, that the universe is tracing out all possible histories, rather than just one privileged one?

Everett was able to demonstrate that, in simple but suggestive cases, the development of the probability waves of Hilbert space tends naturally to give rise to different branches of outcomes whose subsequent histories the evolving wave continues to trace out.

Unfortunately, at the time Everett was writing, the mathematics of decoherence had (inevitably) yet to be properly worked out, and it was not entirely clear why histories that were different should continue to diverge and interact with one another less and less, as is of course the case. This valid problem caused another physicist, Bryce de Witt, to try to advance Everett's theories in a way that in retrospect was unhelpful. It was de Witt who coined the term "many-worlds," and sought a mechanism that would cause different worlds to diverge completely from one another, cleaved apart by outcome lines that had zero probability. We can explain his idea with the version of the two-slit experiment diagrammed in Figure 9-1.

The height of the wave function indicates that the particle involved is more likely to turn up in some places than others, but at some points it can drop to zero; interference cancellation is perfect, and the particle should never be detected in such a position. De Witt tried to interpret such points as fault lines, splitting the universe permanently into distinct versions, each corresponding to one of the possible regions in which the particle might end up. This is neither correct nor necessary. There is no point at which outcome worlds diverge completely. They continue to interfere with one another, although in a way that decreases rapidly with time. But they never actually split.

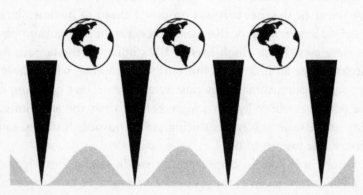

FIGURE 9-1 De Witt viewed zero-probability outcomes as giving rise to segregated worlds, like lane barriers forcing automobiles to diverge toward different destinations at a road junction.

When Everett first developed his theory, he made no reference to splitting worlds. Rather, his theory describes a single universe that processes many different versions of events. A good metaphor for this grander vision of the universe—often called the *multiverse,* to distinguish it from the single version of reality visible to a single version of a single observer—is a type of computer that was proposed during the 1980s. This was an optical computer consisting of bundles of glass fibres and other light-transmitting components, joined in the same kind of arrangement as the wires in an ordinary electrical computer. But the optical computer would be able to do many things at once, simply by shining in slightly different wavelengths of light using appropriately tuned lasers.

To observe the result of a calculation input using, say, a blue laser of wavelength 2,345 Angstroms, you would just use a corresponding blue filter at the far end to screen out all the light bouncing around the system from other users. Thus a single set of hardware could simultaneously process different calculations for different users. For example, rival weather forecasters could use the same hardware at the same time to generate different predictions for the weather. In just the same way, Everett's multiverse-wavefunction simultaneously calculates many versions of what we call reality.

According to Everett, you see a single version of reality because the countless divergent versions of patterns of neuron firings in your brain very rapidly cease to affect one another, just as 2,345-Angstrom calculations in the computer described above are affected only by light very close to that particular wavelength. Other versions of reality—which of course include other versions of your brain—quickly become imperceptible to your own version.

However, thanks to de Witt, the false image of universes actually splitting quickly became associated with many-worlds. Famously, John Wheeler ultimately rejected his pupil Everett's theory as having too much conceptual baggage. Perhaps the notion of the universe repeatedly splitting was the major part of that baggage.

Improbable Numbers of Worlds

The main feature of many-worlds that both physicists and laypersons find disconcerting is the sheer vastness of the multiplicity it implies. Philosophers use the term "ontological extravagance." That is just a grand way of saying what Paul Davies and others have put more pithily: If many-worlds obeys Occam's razor insofar as it is economical in assumptions, it is vastly extravagant in worlds. Is it more sensible to prefer fewer assumptions, or fewer invisible worlds?

In the history of science, however, there are many excellent precedents for accepting economy of assumptions over economy of worlds. Just a few hundred years ago, most astronomers believed that the universe consisted of our own solar system, a single sun orbited by half a dozen planets. The stars seemed mere insignificant pinpricks of light, although their lack of apparent motion as the Earth traced out its billion-kilometer orbit implied that that they were in reality distant and, therefore, bright objects. But then it was noticed that the apparent positions of some stars relative to others does shift slightly, just as would be expected to happen by parallax if they were all at different ranges. Careful measurement enabled the distance to the nearer stars to be calculated. To appear as bright as they do, it turned out that they must be objects quite similar to our own Sun in size and power. They might even possess planets of their own.

The progress did not stop there. About 100 years ago, the universe was thought to consist only of our own galaxy. But scattered among the normal stars, which are pointlike even when viewed through the most powerful telescope, were fuzzier, more extended objects. At first they were assumed to be clouds of dust and gas within our own galaxy, but under closer examination, some of them displayed a pattern of luminosity quite different from that which such a cloud could generate, unless previously unknown physics was involved. The choice was between positing a new law of physics or accepting that we live in an incomparably vaster universe than conceived up to that point, containing a hundred billion galaxies. Many astronomers had great difficulty coming to terms with the latter view, although very few people would doubt it today. We now accept that the universe contains not one sun, but 10^{22}—all this from deductions about tiny points of light, even the nearest of which we may never get to visit.

Accepting the reality of the many worlds of quantum is merely the next step on a ladder we have already learned to climb. The idea that we live in a vast Hilbert space is admittedly startling at first encounter, just as the idea that we live not on a flat plane but on a round lump of rock plunging through a vastness of vacuum was startling when the human race first encountered it.

We can never see those other world lines, with different histories from our own. But here is a parable that might help convince you. Imagine that you are traveling on a ship, and you don a pair of special glasses that let you see a little way into diverging quantum world lines, an extrapolation of the kind of experiment described in Chapter 10. To your astonishment, you see that the ship keeps blurring and then separating into two equally solid-looking copies, which rapidly diverge to left and right. Sometimes you are on the right-hand ship, and sometimes on the left-hand one. You can get only a very brief glimpse of the other ship each time, but you can see yourself on it, and you can just see the events on board beginning to diverge from those of your own vessel before it becomes lost in the mist.

Should you arbitrarily assume that each time a duplication occurs, you always happen by good luck to be on the only ship that is real? Or are the ships you do not happen to be aboard just as entitled to reality? To me, the claim that the other yous are unreal is as silly as those philosophical games in which you are asked to consider that you might be the only real person on an Earth populated with 6 billion cleverly programmed but nonconscious robots. It is a gross violation of the Copernican principle of mediocrity to think that your particular world line must be the uniquely special one every time a divergence occurs.

If only we could do a clear and unambiguous communication-between-worlds experiment. Then there would be no room for argument about the reality of many-worlds. Unfortunately, the laws of physics do not seem to allow such a thing.

This is frustrating because two potentially useful methods of harnessing the power of many-worlds, which we will look at in detail shortly, can be described in terms of sharing resources between worlds, or even sharing information between worlds. For example, a loose way of describing the operation of a quantum computer is as follows: As worlds start to diverge, hundreds of billions of different copies of the computer come into existence. Each of these computer copies can work on a different calculation. The shared result of their labors, however, can be made available to all the diverging worlds created when the bubble of Hilbert space describing the computer is systematically collapsed by measurement at the end of the calculation.

This makes it sound as if Hilbert space might possibly be used as a kind of mailbox for communicating between worlds. Unfortunately, the mathematics that describes Hilbert space rules this out because it implies that everything that goes on in Hilbert space is reversible. As soon as you try to take information out of Hilbert space, that reversibility is destroyed. Such acts of measurement, by definition, cause decoherence. You can preserve multiworld access to a bubble of Hilbert space only by allowing it to evolve undisturbed. It reminds me of C.S. Lewis's "Wood Between the Worlds" described in *The Magician's Nephew*. Any Hilbert space accessible from more than one world line must be a timeless place, in which we can leave no permanent mark.

The Sociological Problem: Fear of Being Misunderstood

Asking prominent physicists whether they really believe in many-worlds is a tricky business. Undoubtedly, one reason why physicists are reluctant to come out as many-worlders is the fear that their views will be misunderstood or caricatured in a science-fictional kind of way: "Tell me, Professor, might we be able to set up a quantum radio link to a world where the South won the American Civil War?" These fears are not groundless. It is a fact that science-fiction writers were exploring the notion of parallel worlds long before Everett came up with his many-worlds perspective on quantum mechanics. A follower of Everett treads a tricky path. If asked, "Are there worlds somewhere out there where the South won the Civil War?" the honest Everettian must reply, "Yes." Explaining why we can never see such a world, or talk to its inhabitants, is a subtler matter.

Yet the fact that we can never visit a place is no grounds to deny its existence. Even in the classical universe, we can see distant galaxies that we can never possibly visit, because even if we were in a rocket traveling just a whisker short of the speed of light, the continuous expansion of space that has been going on ever since the Big Bang would carry them beyond the edge of the portion of the cosmos accessible to us before we got there. Yet we do not doubt that those galaxies are as real as our own. An alien living in such a galaxy would have no fear that he would blink out of existence at the moment he passed out of Earth's sight.

But I do not mean to imply, of course, that all physicists who are reluctant to endorse the many-worlds hypothesis are doing so out of stubbornness or moral cowardice. There are genuine issues still to be resolved, dragons lurking in the undergrowth of many-worlds, and we shall come to them in later chapters. But first let us look to the positive. There are remarkable ways to harness the power of quantum that would be much harder to understand, or for that matter to invent in the first place, without the benefit of many-worlds insight.

CHAPTER 10

HARNESSING MANY-WORLDS 1

Impossible Measurements

The Gap in the Curtains

In the final chapter, we will look at some controversial tests that might prove the correctness of the many-worlds interpretation beyond doubt. But there is one kind of experiment that has already been done successfully and could be said to demonstrate not only that worlds in which history unfolds differently are real, but also that communication between worlds is possible, at least in a carefully defined and limited way.

The basic procedure is known as the Elitzur-Vaidman experiment, after its original proposers. I had the privilege of meeting Lev Vaidman several times when he spent an extended period in Oxford. A small and rather gnomish man, he reminds many people of a younger version of Roger Penrose. But their views on quantum physics could not be more different. Vaidman is a strong supporter of the many-worlds view, and he fascinates his students by proposing highly imaginative thought experiments that more staid academics might dismiss as science fiction.

There is nothing hypothetical about the Elitzur-Vaidman experiment, however: it has now been performed many times, in increasingly sophisticated variants. The basic piece of apparatus involved is something called a Mach-Zender interferometer, illustrated in Figure 10-1. As a tool for discriminating between wavelike and particle-like behavior, it is to the two-slit experiment what a Harley-Davidson is to a pushbike. A beam of light is fired from point O, as shown by the arrow. It encounters at A an optical component called a half-silvered mirror, which reflects half of the light energy upward toward B; the other half carries on toward C. Reflected back together by standard mirrors at B and C, the beams recombine at another half-silvered mirror D, where again half of each is reflected and half transmitted, all the light ultimately reaching detectors E and F.

How many photons end up at E, rather than at F? If photons were classical particles, the answer would be obvious. At each of the two half-silvered mirrors, each photon has an equal chance of being transmitted or reflected. So one-quarter would end up following each of the four routes: ABDE, ABDF, ACDE, ACDF. In the end, half would reach

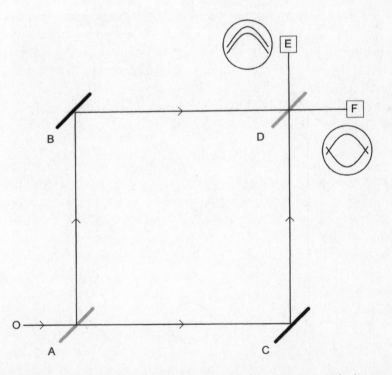

FIGURE 10-1 Mach-Zender interferometer, the perfect wave-or-particle detector.

E and half F. But in reality, photons—even individual photons—also show wavelike behavior. We can arrange the geometry so that the routes ABDE and ACDE are exactly the same length, but the routes ABDF and ACDF differ in length by a small amount, exactly one half-wavelength of the light being used. Now, the detector at F receives no photons because the waves cancel as shown, just as they do at the center of a dark band in the two-slit experiment. All the photons arrive at E.

Now for the clever bit. If, in the two-slit experiment, we close one of the slits, of course the interference pattern disappears. This means that an observer who was initially positioned at the center of one of the dark bands in the interference pattern, and therefore saw no photons at all when both slits were open, now starts to receive some. Something similar is true when you block one of the routes through the Elitzur-Vaidman layout, for example, by placing an obstacle as shown in Figure 10-2.

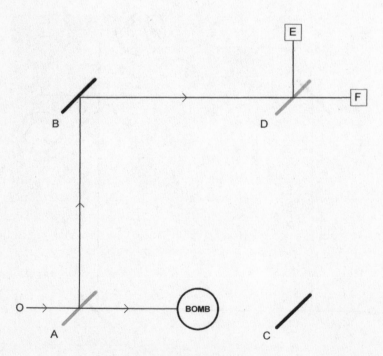

FIGURE 10-2 Mach-Zender interferometer with bomb.

Now half of the photons sent from O try to take the lower route via C and are absorbed by the obstacle. But the remaining half travel safely via B, and then half end up at E and half at F. With no interference to cause complications, they behave like classical particles.

Suppose that you have a setup like this, and you do not know whether the path via C is blocked? Ever the showman, Vaidman dramatizes the situation. Suppose the potential obstacle is a bomb wired up to a photon-detector detonator. Is it possible to test if the detonator is there without setting the bomb off? Extraordinarily, it is feasible to do this. If the detonator is not there, the situation is that of Figure 10-1: A photon fired into the apparatus always ends up at point E, and the detector at F never registers. If, however, the detonator is present, as in Figure 10-2, a photon fired in from O has a 50 percent chance of continuing toward C and setting off the bomb. But if the photon is instead reflected via B, it then has a further 50/50 chance of ending up at E or F. Each photon we fire therefore has a one in four chance of registering at F, warning us that the detonator is there *without* setting the bomb off.

How can this have happened? How can a photon that never went near the detonator tell us whether it is present? It is tempting to think in terms of some kind of prober waves or guide waves that must have done the job. But these would of course correspond exactly to Bohm's pilot waves. As we saw earlier, there are almost insuperable problems with this concept, including pathologically nonlocal behavior. If we think in terms of interfering many-worlds, however, there is a far simpler explanation. Whenever a photon hits the half-silvered mirror A, two worlds are effectively created. In one of them, the photon continues toward C. In the other, the photon is reflected upward via B. These worlds continue to interfere—until a photon measurement is made in either of them. Suppose the photon in our world happens to go via B. It continues to be affected by its counterpart in the parallel world that went via C—but only up to the point where the counterpart is measured by being absorbed. If the path via C is clear, this results in interference that prevents a photon from being detected at F, as in Figure 10-1. But if the photon following path C is measured by striking the bomb detonator, as in Figure 10-2, the link between our world and the parallel one is disrupted at that point; it has no further effect on our own. Interference ceases, and it is possible for our own photon to hit F.

How has the trick, which alarmingly resembles the communication of information between worlds, been accomplished? The absence of a signal can contribute information, like the famous example of the dog in the Sherlock Holmes story that did not bark in the night. The present situation is more like a general who sends a scout to see if the enemy is hiding behind the next hill. "If all is clear, detonate this green flare," he tells the scout. He does not need to give him a red flare to signal the presence of the enemy, for in that case the scout will be dead: The mere absence of a green signal at the prearranged time will tell the general all that he needs to know, one bit of information. The spooky thing about analogous quantum measurements is that we are using a signal not from another hilltop, but from another world. If an "OK" interference signal does not come, our scout—our otherworldly shadow photon—has fallen out of communication.

The Elitzur-Vaidman bomb detector is not very efficient: It is twice as likely to set the bomb off as it is to give a useful warning. It is ironic that a much more effective method has been devised and demonstrated by one of the arch-opponents of many-worlds, Anton Zeilinger.[1] It uses the basic setup shown in Figure 10-3.

The core of the device is a racetrack, with a mirror at each corner, round which a photon can circulate many times. There is a switching system, S, by means of which a photon can be introduced into the system, and extracted at a chosen later time. At R is an optical component called a polarization rotator. It turns the polarization of every photon that passes through it by a fixed amount, say, one degree clockwise. If we introduce a vertically polarized photon into the system, allow it to circle 90 times, then extract and measure it, we will find it is now horizontally polarized.

So far, so obvious. But now we introduce an alternative path into the system, as shown in Figure 10-4.

The mirrors K and L are more sophisticated variants of the half-silvered mirrors used in the previous bomb tester. They have the property of allowing vertically polarized photons to pass unhindered, whereas horizontally polarized photons are always reflected. So any

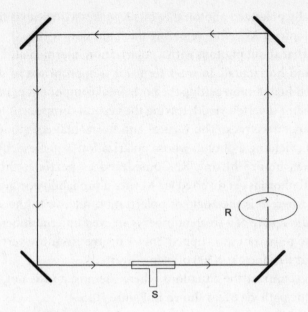

FIGURE 10-3 Zeilinger racetrack.

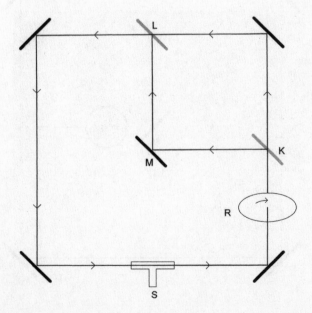

FIGURE 10-4 Zeilinger bomb tester.

horizontally polarized photon that hits K makes a dogleg via the conventional mirror M before rejoining the main flow at L.

But what about photons with a polarization intermediate between vertical and horizontal? In wave terms, it is appropriate to think of each individual photon getting the horizontal component of its polarization vector diverted via M, leaving the vertical component to travel via the outer racetrack. The vertical and horizontal components reunite at L, yielding a photon whose polarization is just exactly whatever it was before hitting K. Considered as particles, however, individual photons get diverted via M with a probability proportional to $\sin^2 a$, where *a* is the angle of polarization relative to the vertical. Because the square of a small number is an even tinier number, a photon whose polarization is tipped only 1 degree from the vertical has only about 1 chance in 3,300 of being diverted.

Let us contrast the situations where there is, and is not, an obstacle in the path via M, as shown in Figure 10-5.

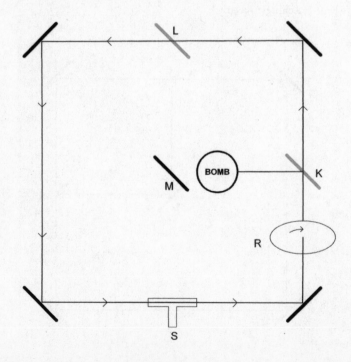

FIGURE 10-5 Zeilinger bomb tester with bomb.

If there is no obstacle at M, the diverting mirrors have no net effect. A photon that has a nonvertical polarization before hitting K has
that same nonvertical polarization after leaving L. So the photon will
still rotate polarization 90 degrees after 90 transits, just as in the basic
setup of Figure 10-3. The presence of the mirrors K, L, and M makes
no difference. But what if we introduce an obstacle into the route via
M? Now the horizontally polarized component of each photon created at K gets absorbed, never reaching L. The photon will go round
and round the track, knocked 1 degree from the vertical each time it
hits R, but restored to the vertical at K, and remaining vertical at L.
Extracted after 90 transits, it will still be vertically polarized. If we get
back a photon that is vertically as opposed to horizontally polarized, it
therefore warns us: Beware, there is a bomb.

Now for the extraordinary bit. Because the photon considered as a
particle has only about 1 chance in 3,300 of being diverted via M on
each circuit, the chance that it has gone this way during any of its 90
circuits is still only about 1 in 37, and the bomb is correspondingly
unlikely to detonate. We have achieved something even more impressive than exchanging information between one world and another. We
have in some sense communicated a bomb warning from a small set
of worlds where the bomb detonated to a set 36 times larger that remains safe.

In principle, this could be increased to any ratio we wanted; for
example, to double it, we just reduce the power of the polarization
rotator to one-quarter degree per circuit and allow the test photon to
circulate 360 times. Like the general who sacrifices one scout to protect the rest of his army, we can sacrifice a small number of worlds to
save many others. Of course the chance that your world will be the one
in which the bomb goes off never quite shrinks to zero—just as however large the general's army, there is always a chance that you will be
picked to be the scout.

It is this ability to share information profitably between worlds—
to export information generated in one world to a potentially unlimited number of others—that, in the view of David Deutsch and his
colleagues, will open up the extraordinary potential of quantum computers. Although Anton Zeilinger sees things differently, I once heard

him make a practical point about experimental design that could be interpreted almost poetically from the many-worlds viewpoint. He pointed out that quantum spookiness becomes most apparent when we measure things at small angles. Probabilities we might expect to be proportional to a are instead proportional to the much smaller quantity a^2. This was true of the lottery cards in Chapter 1. If the spot color changes from black to white at some place on a 90-degree arc and they were classical cards, then two marks scratched 6 degrees apart should have had a 1-in-15 chance of being a different color, but spooky quantum effects reduced this to nearer 1 in 100.

We have just seen that small measurement angles are similarly the key to efficient quantum bomb detection. Long ago I read a story by John Buchan, called "The Gap in the Curtains," about an attempt to foresee the future. Peeking through gaps at narrow angles turns out to be, in sober fact, the way to peer between the curtains that normally hide parallel worlds from our sight.

Lifting the Veil

A third, and even spookier, type of bomb detector is shown in Figure 10-6.

The central object is a block of transparent material resembling a large gemstone called a monolithic resonator. This is an intimidating name for a very simple device whose key property is that it can, in principle, trap light in an endlessly circulating path. If the two triangular prisms at the bottom of the diagram were removed, then a photon circulating within the octagonal block would never be able to escape, because the refractive index is high enough that total internal reflection occurs in turn at each of the points A, B, C, and D. Of course the photon does not really circulate forever, because the block can never be made perfectly transparent, but an average photon lifetime of thousands of circuits is perfectly possible.

If, however, we bring two triangular prisms up to almost touch the resonator at points A and B, as shown, total internal reflection at these points is now said to be frustrated. As a photon bounces round and round within the monolith, it has a small chance of escaping at

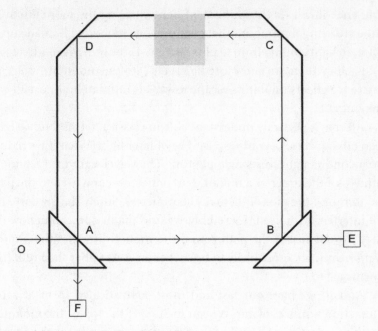

FIGURE 10-6 Monolithic bomb detector.

either of these corners. Conversely, we also have a way of injecting photons into the monolith, for example, from O. The behavior of the system turns out to be most interesting if we adjust the tiny gaps between the prism and the monolith so that, under normal circumstances, reflection is much more likely than transmission. Then a photon fired in from O, behaving in a particle-like way, is most likely (say, 99.9% probable) to get reflected straight down to F, without ever entering the monolith at all. On the other hand the occasional photon that does get into the monolith will typically circle a few hundred times before escaping at either E or F.

However, this scenario ignores the wavelike properties of light. Suppose we spray into O a continuous wave of light using a laser, for example. Now we can expect interference; the whisper of light that enters the monolith and is reflected round the path A, B, C, D, and back to A has a chance to interfere with a later portion of the wave. If we make the path lengths right, we can arrange that constructive interference increases the amount of light entering the monolith at A, while reducing the amount that goes downward toward F (made up of reflected light coming from O plus straight-through light coming from D). Each cycle, more and more light gets into the monolith; since interference effects inhibit its escape toward F, ultimately almost all will leak out at E.

All this is perfectly understandable in classical terms, but we have been talking about a continuous wave of laser light. What if we reduce the incoming light to a single photon? The wavelength of the photon is tiny—of the order of a millionth of a meter—compared to the path length round the monolith, many centimeters. Surely the photon cannot interfere with itself? Incredibly, we find that it does. Somehow the mere *availability* of the path round the monolith makes the photon overwhelmingly more likely to be sucked in at A, rather than reflected downward to F.

And so we have our last and most sophisticated form of zero-interaction bomb detector. We can include at the top of the monolith a bath of transparent liquid of the same refractive index as the glass of the monolith—completely invisible to the eye, although I have drawn it faintly shaded to help us see what is going on. If the path round the monolith is blocked, as in Figure 10-7, a photon fired in from O behaves in a particle-like manner, and is almost certain to be reflected straight down into F, without entering the monolith or setting off the bomb. But if the path round the monolith is clear, as in Figure 10-6, the mere *possibility* that the photon can go round the path as many times as it likes is enough to ensure wave like behavior: The photon is almost certain to be sucked into the monolith, and eventually detected at E.

This is what the math of quantum mechanics predicts, but surely it is too bizarre to be explained in intuitive terms? Actually, it can be explained quite well even in terms of the guide waves of Chapter 2. In the surfer-and-guide-wave picture, we must think of the surfer as occupying a position that is uncertain not merely in the sense of not knowing where on one particular wave front he is, but also in the sense of not knowing on which of a series of possible wavefronts he is riding. This new kind of guide wave consisting of a whole series of waves is

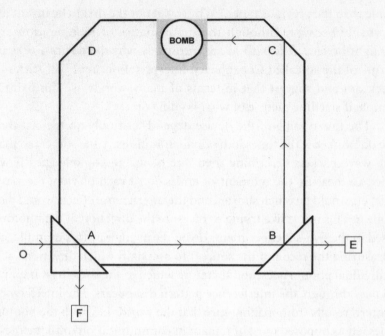

FIGURE 10-7 Monolithic bomb detector with bomb.

called a wave packet, and is illustrated from the side in Figure 10-8; on measurement, the photon will be found occupying some particular position within the packet, as indicated by the denser shading. The photon may be tiny, but the guide-wave packet can and does interfere with itself.

The monolithic detector, described in a brilliant 1997 paper by Harry Paul and Mladen Pavicic, is much more efficient and practi-

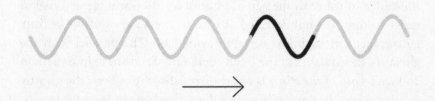

FIGURE 10-8 Photon wave packet.

cable than the previous types.[2] Why was it not the first to be invented? Probably because, although the mathematics of its operation are straightforward, it is hard to see intuitively why the device works in terms of the so-called Copenhagen interpretation. Yet I will stick my neck out and suggest that in terms of many-worlds, we can paint a simple if startling picture of what is going on.

The functioning of the device depends critically on the fact that we do not know the times individual photons leave the source, so that the wave packets describing them are of macroscopic length. If we tried to measure the moment of emission of each photon, the wave packets would be much shorter, and the interference effects would disappear. This is just like trying to measure the direction of the photons used in the two-slit experiment. If you try in any way (for example, by measuring the recoil of the source) to ascertain which direction each individual photon goes, and therefore which of the two slits it is going to pass through, the interference pattern disappears. The interference pattern results from making sure that the worlds in which the photon goes left as opposed to right remain in communication until the measurement on the photographic plate is made.

In the case of the monolithic detector, however, the worlds that must remain in communication are not those in which the photon went left or right, but those in which it left the source earlier rather than later. From the point of view of a world in which we get a click at detector E, and thus know in an almost risk-free way that there is no bomb present, there are ghost worlds in which the photon left the source earlier, raced once around the monolith, passing through the bath of liquid at the top, and then effectively beckoned subsequent ghost photons in, until a growing horde of ghosts that had already been round the monolith once, twice, thrice, and so on acquire enough substance to usher in the actual photon of the world we perceive as real—so that it is not detected at F, as in the absence of the ghostly encouragement it almost certainly would be. The parallel with the ghosts of the Marshes of the Dead beckoning Frodo in to join them in Tolkien's *Lord of the Rings* is almost irresistible! But here, the ghostly scouts beckon the photon in only if it is indeed safe to enter the monolith.

You might feel that this latest example of many-worlds effects is even spookier than those earlier in this chapter. And you are right, because we are now making use of a world that is in a sense ahead of our own in time, a world in which the photon will already have triggered the bomb if it is present. The importance of the monolithic reflector is that it delays a photon by trapping it, unmeasured, for a significant period—thus preserving communication with that other world. In present-day apparatus the time lag involved is only a few nanoseconds, corresponding to a wave train a few meters long, but in principle this time could be greatly extended. You might be making use of information from worlds where, if a real bomb had been present, you would already have been dead. What are we to make of this?

I would suggest that it might well throw light on a puzzle we have already touched on: the alarming phenomenon of particles that appear to quantum tunnel faster than the speed of light. This is analogous to thinking that the photon in Figure 10-7 must have gone faster than the speed of light in order to have had time to explore the region of space that contains the bomb. The truth is subtler: We are making use of information from other-worldly variants of the photon that traveled no faster than light, but simply left the source earlier. Similarly in the quantum tunneling case, as long as the "tunneling" particle is still in flight we remain equally in touch with worlds where it departed the source earlier and where it departed the source later. The key insight is this: The fact that interaction in *either* world causes the link to collapse prevents any faster-than-light messages from being sent via such particles, even though they might be effectively displaced in time in different worlds.

Zero-interaction measurement devices might well be capable of practical applications. How wonderful it would be if we could, for example, take an X-ray of a pregnant woman without the usual danger of damage to the fetus from high-energy photons—because the photons making the photograph, or at least the vast majority of them, did not pass through her body at all. Although medical applications are still some way off, interaction-free measurement and testing in other contexts might well be realistic. Adrian Kent and David Wallace have recently coauthored a paper describing a kind of testing device based on the principle.[3]

It is truly ironic that one often hears statements like, "In a quantum world, you cannot measure any object without affecting it at least slightly." That is the precise opposite of the truth. In a classical world, we would really not be able to measure anything without affecting it, because every photon or electron would have some effect on whatever it struck, however gently. Only in a quantum world does it become possible to measure something without affecting it at all.

CHAPTER 11

HARNESSING MANY-WORLDS 2
Impossible Computers

There is a sense in which quantum computers represent the triumph of the many-worlds interpretation to date. Not because the feasibility of quantum computers proves the reality of parallel worlds—that claim is hugely controversial, and we will scrutinize it later in this chapter. But what is inarguable is that the many-worlds viewpoint helped David Deutsch, back in 1985, have the key insight that made quantum computing possible.

Deutsch was not the first person to speculate about the possibility of quantum computing. A couple of years earlier, Richard Feynman had already published a paper on the subject. [1] The ever-imaginative Feynman put forward a whole range of ideas for harnessing quantum to make computers smaller and more powerful. However, his real interest was in making computations of a type that I would call analog rather than digital.

The idea behind an analog computer is that a physical quantity, for example, the flow of electric current or the amount of charge on a capacitor, can be used to represent a numerical quantity to high precision. For example, if you can measure the charge or current to one part per thousand, you can use it to represent a quantity to about three decimal digits of precision. If you can improve the accuracy to one in a million, you get six decimal digits.

At first sight, this seems far more efficient than a digital computer, where the current or charge in a given component is allowed to have one of only two distinguishable values, on or off, representing a single binary bit. The advantage does not end there. You can build analog electrical components to add, multiply, and divide such currents in a single hardware operation, a job requiring hundreds of individual bit-operations in a digital computer. In a nonquantized world, there is in principle no limit to the amount of information you can store in a single analog quantity.

It occurred to Feynman that even in our quantized world, systems that can yield only a limited amount of information on measurement—like which one of two detectors a photon or electron ends up hitting—may nevertheless signal the outcome of an enormous amount of behind-the-scenes processing. In single-world terminology this processing is effectively done by the evolution of a probability wave, which can develop a very complex form and can be made to interfere with itself in intricate ways. Feynman realized that you could in principle make the probability wave do a large amount of analog computation before registering its simple zero-or-one outcome.

The main application he envisaged was the simulation of quantum systems themselves. In the macroscopic world, the behavior of a small physical model can be used as a kind of analog computer. For example, when a model of an airplane is placed in a wind tunnel and the fan is started, you are effectively using this system as a computer to calculate the forces on something quite different, a full-sized airliner moving through the real atmosphere at a much higher speed. Feynman thought that analogously, small and relatively easily controlled quantum systems might be used to predict the behavior of much larger and more intractable ones.

Unfortunately there is a fundamental problem with analog computing—the impossibility of achieving true precision. Whether you are working with macroscopic currents or quantum probability waves, in practice you can never make the amplitudes have exactly the values you wish, to infinite precision. Moreover, in most types of computation you might want to perform, any tiny initial errors rapidly multiply. In the early 1980s, I was occupied writing algorithms to verify results obtained on a classical analog computer that could at that date still do certain calculations much faster than its digital cousins—but the analog computer never gave exactly the same result twice. I vividly remember that the analog machine gave different results on a warm day than on a cold one, despite being installed in a supposedly temperature-controlled room. Traditional analog computers simply cannot produce reliably repeatable answers and they have duly fallen out of use. Attempts to use quantum probability waves to calculate analog quantities would suffer the same problems.

David Deutsch's many-worlds perspective led him to favor a subtly different approach. He could see the potential of something much more like a conventional digital computer, but one in which a single set of hardware could perform an enormous number of different computations at once—as he sees it, different versions of the computer in a sense simultaneously performing different calculations in different worlds. Using this idea, a relatively simple machine could do an enormous amount of processing. He called the technique "massive parallelism." This concept, which he described in detail in a landmark 1985 paper, is the basis of all modern designs for quantum computers. [2]

Parallel Computing

The real temptation of the approach is the sheer number of computers you could in effect generate. The idea of parallel computing, using a large batch of small computers working together to solve a problem, is not new. As I write, the world's largest parallel-processing supercomputer is Japan's Earth Simulator, with 5,000 individual processors collectively capable of some 40 trillion calculations per second. Five thousand processors is by no means the limit, however. Ingenious scientists have found a few applications that allow thousands or even millions of processors to work on the same problem with very little communication required between them. This allows vast numbers of desktop computers to take part over the Internet by running screensaver programs—programs that work when the computer is otherwise idle, and normally just generate pretty patterns to amuse the user. Provided you have a computer that at least occasionally connects to the Internet, so that the program can download data to work on and upload the results, you (or anyone else) can take part in a variety of worthwhile projects.

The most famous of these is the SETI@home project to process huge amounts of radio telescope data, to see if any signal from any part of the sky shows a pattern that might indicate that it is an intelligent signal from an alien race. The SETI project has already received plenty of publicity, but there are now a number of other distributed-screensaver projects to choose from. If you want to do something useful with your computer's spare moments you might like to try the Screensaver Lifesaver Web site.[3] This project involves screening millions of possible chemicals for useful biomedical effects, in particular against cancer, by calculating to what extent their shapes will cause them to dock with certain target molecules. It's an extremely worthwhile cause and is likely to be the forerunner of an even more ambitious project as the Human Genome Project gives way to what has been called the Human Proteome Project, the huge task of identifying every biologically active molecule in the human body. Another interesting option is the Climate Prediction screensaver.[4] This builds on the technique of ensemble weather forecasting. Nowadays, when a weather forecasting center makes a prediction, it is never the result of a single computer run. Forecasters are aware of the possibility of the butterfly effect: Would a tiny change in input parameters have produced a completely different forecast? They therefore run their model many times, each time with slightly different starting values because they know that their input data can never be perfectly accurate. If they get the same outcome every time, they know that they can forecast that weather with high confidence. If two or more significantly different results come from different runs, they know that they cannot be so sure. When a forecaster says that there is 33 percent chance of rain tomorrow, he may well be indicating that of 100 such simulations, about one-third ended in wet weather, the rest in dry.

The Climate Prediction screensaver takes the technique a bit further. When predicting what the climate will be like in 20 or 50 years, there are an enormous number of unknown variables governing feedback effects. How much sunlight will be reflected back into space by clouds in a given scenario? By what percentage would a 1°C temperature rise increase the melt rate of the Greenland ice shelf? How much will the Gulf Stream weaken per point of decrease in the salinity of the Northern Atlantic? And a thousand similar questions. The Climate Prediction project is building not a single forecast, but an ensemble of likelihoods. They already have one interesting result: A kind of butterfly effect affects the climate, not just the weather on some particular day. A very tiny change in parameters can give a whole region a markedly different temperature and rainfall over a long period. The Climate Prediction screensaver is beautiful and somewhat terrifying to watch, as a graphic of Earth shows the icecaps, deserts and forests shrink and grow in your particular slice of the future.

These kinds of distributed Internet projects might be able to recruit up to several million computers each. But obviously there is an upper limit. Even if you could persuade everyone on the planet to help you, there are only a billion or so computers available. That is nothing compared to the potential richness offered by quantum.

Massive Parallelism

To understand the benefits of quantum computing, we will briefly review how a standard computer works. The heart of a computer is a device much like an old fashioned mechanical calculator, the kind that had a handle on the side that you could turn to add, subtract, or multiply numbers using wheels very similar to those in the odometer on your car dashboard. But a computer operates electrically, and it uses numbers based not on the decimal system, which has 10 different digits with successive columns representing units, tens, hundreds, and so on, but on the binary system, which has just two digits, zero and one, and in which successive columns represent units, twos, fours, eights, etc. Figure 11-1 illustrates a simple binary calculation.

Deutsch's initial idea amounts to this. Suppose we could take a perfectly ordinary binary calculator and place it in Hilbert space, enclosed behind an impenetrable barrier like Schrödinger's cat. We could then arrange the digits poised in unknown states, so that just as the cat's Hilbert space would explore both the live-cat and dead-cat possi-

Translating decimal into binary is easy:

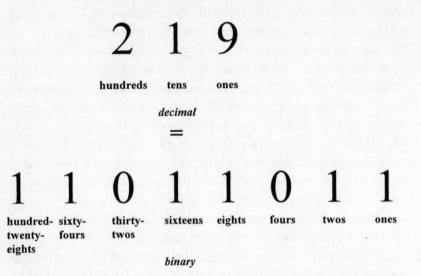

It is also straightforward to do arithmetic in binary. For example the two sums below are equivalent:

$$
\begin{array}{r}
2\ 1 \\
+\quad 2\ 7 \\
\hline
=\quad 4\ 8
\end{array}
\qquad\qquad
\begin{array}{r}
1\ 0\ 1\ 0\ 1 \\
+\quad 1\ 1\ 0\ 1\ 1 \\
\hline
=\quad 1\ 1\ 0\ 0\ 0\ 0
\end{array}
$$

FIGURE 11-1 Binary numbers and arithmetic.

bilities, the calculator would explore a calculation in which the digits were in every possible combination of zeros and ones. If you imagine the computer sitting in a small cubicle with a human operator, then creating the impermeable barrier that separates the system from the outside world effectively creates a huge array of cubicles. I shall call this array the Dilbert Hotel, with apologies both to Scott Adams, creator of the excellent Dilbert cartoons, and to mathematicians familiar with the Hilbert Hotel, setting for David Hilbert's famous thought experiments to illustrate the possibilities of infinities.

How many rooms has the Dilbert Hotel? Well, an 8-digit binary register can be set to $2^8 = 256$ possible combinations. So if we are multiplying every possible 8-bit number by every other possible 8-bit number, we are performing $256 \times 256 = 2^{16}$ calculations in a hotel with 65,536 rooms. That might not sound vastly impressive, but modern computers have 32-bit or 64-bit registers. A 32-bit register can be set to just over 4 billion different combinations. In multiplying every possible 32-bit number by every other 32-bit number, we generate a hotel with more than 16,000,000,000,000,000,000 (16 billion billion) rooms—far more than the total number of computers ever built. This is beginning to sound promising.

However, the problem with this image is that it leads us to expect a great deal too much. You can whimsically imagine the operator in every cubicle doing his own thing, following a different line of thought, like in a real office block housing billions of computers and programmers. Unfortunately the reality is far more mundane. Each cubicle differs from its immediate neighbors by the setting of only one binary digit, and each worker must respond to the same sequence of commands shouted over some public announcement system. The workers are mannequins, all jerking about to the same string-pulls, as if following the steps of a formal and intricate dance.

A more accurate visualization would be to take a single cubicle and equip it with floors, walls, and ceilings that are perfect mirrors, thereby creating an illusion of a vast number of extra cubicles stretching off to right and left, upward and downward. In a sense we have created something extra—each extra cubicle is visible from a slightly different angle, so to speak—but we certainly have not created billions of independent worlds.

A major practical limitation is that collapsing the hotel at the end of the calculation leaves just one cubicle selected at random from the original array. Consider the task of dividing a very large number into its prime factors, a problem that arises in code breaking. You might naively think that one way of using the Dilbert Hotel would be to have every operator try dividing a different number into the original. Soon, in just one of the vast array of cubicles a lucky Dilbert will be waving his arms over his head, shouting "I have cracked it! The remainder is zero, the number I was given to try is the answer to the problem!" But the chance that we will just happen to get that particular cubicle when we collapse the system is negligible.

To get a useful result from the Dilbert Hotel, we must arrange that every cubicle will hold a copy of the answer that we seek. That is possible, because the Dilberts are allowed to exchange information—to sneak notes between one another over the cubicle walls, so to speak. But they are only allowed to pass on such information (actually by interference effects) in a synchronized and stylized way, all moving to an invisible drumbeat. Then collapsing the system by a measurement at the right moment will give us a Dilbert who is certain, or at least reasonably probable, to be holding the correct answer. The Dilberts are, however, further constrained because time's arrow must not be allowed to operate. For the cubicles to remain in contact with one another, no permanent recording of information can take place. It is as if each Dilbert is not allowed to write anything down, but merely to twist dials to and fro. So commanding him to do something can easily scramble an already useful result that he has found.

With all these restrictions, it is almost surprising to learn that it is, in principle, possible to arrange the rules of the dance so that the final position in every cubicle reflects the answer we are after. The problem is that the method is task dependent; it is entwined with the nature of the particular problem we are using the quantum computer to solve. For each different problem, an algorithm must be found that includes this last information-dissemination stage. This has turned out to be incredibly hard.

A Solution in Search of a Problem

The largest computers have always been needed for just two kinds of jobs, simulating the physical world and code breaking. Indeed, the first really big programmable calculating machines were built for just these purposes during World War II: In Britain, Colossus and its relatives at

Bletchley Park cracked the German Enigma code, while in America, computers at Los Alamos worked out exactly how to ignite a nuclear explosion. Since then, computers have been turned to all kinds of work. Many useful tasks can now be done on relatively tiny and inexpensive desktop computers, but the processing power ideally required for these two heavy-duty applications is still not available. In the case of code breaking, this is because of the arms-race element. More-powerful computers allow more-powerful codes to be generated in the first place. In the case of physical simulations, it is because of the inordinate complexity of the real world. Perfect simulation of anything beyond a medium-sized molecule is still beyond today's computers. The best we can hope for with macroscopic systems like the weather is to achieve ever better approximations.

So, can we use a quantum computer to crack codes or perform physical simulations?

Shor's Algorithm

So far, in 20 years of searching by some of the world's cleverest mathematicians, just one quantum algorithm fully capable of addressing a useful problem has been discovered.[5] And even that case requires a slightly liberal definition of the term useful. The problem involves code breaking.

Nowadays, we all take the convenience of paying for goods and services by plastic for granted. Indeed, even physical plastic is no longer required—you can simply give your credit card details over the telephone, or type them into a form on a Web site. Yet I am old enough to remember the days when almost everything had to be paid for in cash. Even trying to pay for groceries by check earned you a suspicious look from the store manager, and often a surcharge. The transformation has come about largely because of a clever encoding technique that allows an organization such as a bank or a chain of stores to openly publish a key for sending it messages, which cannot be decoded or interfered with unless you have a second key that is kept secret. The point is not that your credit card details are particularly secret—many people, such as waiters, have plenty of opportunity to copy them, and it does not take an Internet hacker. The point is that the transaction details cannot be falsified, so the recipient of such an electronic payment cannot be disguised. A fraudulent debit on your credit card can be traced to whomever it was paid to. The details need not concern us here; the important point is that the cipher system depends on the fact that it is much easier to multiply numbers than to divide them. Thus if I give a standard computer two large prime numbers to multiply, it can do so in a short time. But the reverse task—given the product, which two prime numbers divide it?—would take the computer thousands of years. The difficulty of factoring very large numbers is crucial to the security of modern commercial methods.

Peter Shor realized that a quantum computer could be made to perform this task of factoring very large numbers. His method is so clever that he was awarded the Fields Medal, then the nearest mathematical equivalent to the Nobel Prize, but we are not going to go into the details. The key step was to link the factoring task to the problem of finding the period of a function. The period simply means the interval in which the graph of a regular function repeats itself. For simple functions like sine waves, the period is immediately obvious to the human eye, but it can be much more difficult to spot with sharply discontinuous functions: Imagine a pattern like that of Figure 11-2 but extending for millions of miles. Using Shor's method, factoring a large number becomes equivalent to spotting the periodicity hidden in a really enormous set of random-looking numbers, and this turns

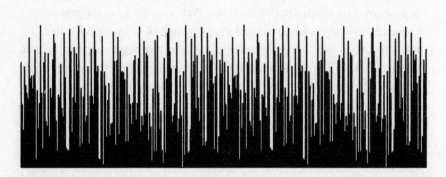

FIGURE 11-2 A discontinuous function.

out to be something a quantum computer can do while staying within the rules of Dilbert Space.

If we could build a computer to run Shor's algorithm, how useful would it be? Really secure messages, top-secret military and diplomatic communications, do not depend solely on these hard-to-factor products of primes. The main reason is that no one has ever proved that there is not a clever mathematical algorithm that could enable large numbers to be factored rapidly on a perfectly ordinary computer. The major use of the prime-number systems is to secure banking transactions.

The only significant result of quantum computers becoming available would therefore be to cause a meltdown of the developed world's economy. That certainly stretches the meaning of the word useful. I am reminded of the tension that arises in commercial companies when a techie announces that he has discovered an interesting problem. He thinks it fascinating, but it probably represents a nightmare for everybody else, who find it interesting more in the sense of the ancient Chinese curse: "May you live in interesting times!"

Hope for the Future

It is a real pity that no one has yet found a good use for quantum computers, because—as so often in the world of computing—it is the software that has turned out to be far harder to implement than the hardware. When Deutsch first worked out the ground rules, quantum computers were pure science fiction. Since then, several techniques have been developed that can perform all the basic hardware functions needed, holding nontrivial numbers of bits in Hilbert space—in other words, so that they interact only with each other without being measured by the outside world. The most promising technology involves supercooled atoms suspended in electric fields. Other problems that Deutsch and others have solved include devising useful computer instruction sets that operate without erasing information—essential to keep the Hilbert space together—and even methods of detecting and correcting errors without prematurely reading the values that must remain hidden until the computation is complete. The latter task is tremendously hard.

As seen from the perspective of a single world, a quantum computer operates not on ordinary binary digits, called bits, but on more subtle entities called qubits. Each memory location contains not a 0 or a 1, but information that can be described by a vector, an arrow pointing to some point on the surface of the Bloch sphere shown on page 89. Thus a qubit can be physically stored in the polarization of a photon, or in the spin of an electron or a larger particle. But the whole point about a qubit is that you do not know where the vector is pointing. This absence of knowledge is what keeps it entangled with the other qubits in the computer, generating the multitude-of-Dilbert-cubicles effect, which can be thought of as a little bubble of Hilbert space. You are not allowed to read the qubit, and something called the quantum no-cloning theorem says that it is also not possible to duplicate the qubit, even without looking at it.

How can you possibly tell if some unwanted interaction with the environment, as is bound to happen occasionally, has corrupted the qubit's value? The detailed answer is too technical to give here, but the analogous method for ordinary computers is shown in Figure 11-3. Without either copying or reading out the values of the central square of bits we can generate an extra row and column of information that enables us to correct single-bit errors. The corresponding technique for qubits is significantly more complicated, but has now been perfected.

So the architecture and hardware challenges of building a quantum computer are well on the way to being solved. What hope is there for the software problem?

I remain optimistic that quantum computers may turn out to have wonderful uses. In the early days of the laser, there seemed to be a very similar dearth of useful applications. Now optoelectronic devices based on lasers are ubiquitous in consumer gadgets, communications networks, and a host of other civilian and military applications. One hope is that quantum computers will be able to solve a very famous class of problems that mathematicians call NP-complete. This is a group of puzzles whose solution time grows exponentially as the system involved gets larger. The most famous is the traveling salesman problem: working out how to visit a group of towns with the least

Data in a conventional computer is safeguarded by the use of checksums. For example for the square of data below, the computer can generate extra digits recording whether the number of digits in each row and column is odd or even:

$$
\begin{array}{cccc c}
1 & 0 & 0 & 1 & \quad 0 \\
1 & 0 & 1 & 0 & \quad 0 \\
1 & 0 & 1 & 1 & \quad 1 \\
1 & 0 & 0 & 1 & \quad 0 \\
\end{array}
$$

$$0 \; 0 \; 0 \; 1$$

If a bit becomes corrupt, the checksums for both its row and column will be incorrect:

$$
\begin{array}{cccc c}
1 & 0 & 0 & 1 & \quad 0 \\
1 & \textit{1} & 1 & 0 & \quad \Leftarrow \textbf{\textit{0}} \\
1 & 0 & 1 & 1 & \quad 1 \\
1 & 0 & 0 & 1 & \quad 0 \\
\end{array}
$$

$$\Uparrow$$

$$0 \; \textbf{\textit{0}} \; 0 \; 1$$

and so the computer knows which bit must be flipped to the opposite value to correct the situation. A more sophisticated version of this technique allows qubits in a quantum computer to be corrected without 'collapsing' the calculation by reading their values to the outside world.

FIGURE 11-3 Error correction.

total travel time. A fast method of solving problems like this would be of immense value to operational researchers, with practical applications in many fields. But it has not been proved that quantum computers can do this.

It remains possible that the "killer app" for quantum computers will involve simulations of quantum systems themselves. It can be tremendously time-consuming to calculate the evolving waveform of even a simple quantum system on a classical computer, and the impossibility of calculating the quantum behavior of more complex entities is becoming an increasing vexation to materials scientists, among others. If quantum computers can be used to understand the behavior of quantum matter, we will have come full circle to Feynman's original hope, but using digital rather than analog computation.

To give a flavor of what the future could hold, remember the cold fusion fiasco when Fleischmann and Pons claimed to be generating fusion power from a lump of palladium in a test tube of heavy water. Many reputable experimenters failed to repeat the result, and it was probably spurious. But the real lesson we should remember is that theorists could not dismiss the possibility that fusion was occurring, because the behavior of real solid matter at the nanoscale—where quantum effects become significant—remains far too complex for today's computers to model. If we get working quantum computers, we might be able to get real insights into what is called condensed matter physics.

Quantum computers with viable architecture and hardware, capable of working with significant numbers of bits, will probably be with us soon. The example of Shor's algorithm proves that they have the potential to be useful. What we vitally need, as ever in quantum, is better thinking tools that will make it more straightforward for humans to program them, to visualize what is going on in the little bubble of the multiverse in which they do their work.

CHAPTER 12

MANY-WORLDS HEROES AND DRAGONS

As we have seen, the battle between proponents of different quantum interpretations has raged for the best part of a century. To my great delight, it is Oxford that has served as the champions' arena for the latest, and I believe probably last, stages of the debate. Oxford is home to David Deutsch, principal champion of the many-worlders, and Roger Penrose, internationally famous defender of the classic single-world view. The two principal devisers of experiments to test the foundations of quantum, Anton Zeilinger and Lev Vaidman, have spent extended periods in town as guests of the University. Oxford's trailblazing Centre for Quantum Computation—now in a sense a victim of its own success, for after an influx of funding it has become a joint Oxford and Cambridge facility, and many new quantum computing centers are springing up worldwide—has attracted researchers whose interest included the practical as well as the theoretical. And so it has been that at conferences and seminars in Oxford, and down the road in London, all the above and many other leading figures have come to speak and defend their views, and to be subjected to polite yet probing questions by their fellow physicists and philosophers of physics such as Simon Saunders, Harvey Brown, and Jeremy Butterfield.

I have given Penrose and Zeilinger chapters of their own. In this chapter I want to focus on the remaining difficulties of many-worlds: How is it that such committed many-worlders as Deutsch and Vaidman, who might seem to outsiders to share extremely similar beliefs, can both describe themselves as in fundamental disagreement about the basic assumptions of the theory? Are the differences as deep as they seem? How much remains to be resolved?

Counting Worlds

There is one acknowledged problem lurking at the heart of many-worlds. It has to do with the relative probability of different quantum outcomes, and the world lines that follow from them.

In simple illustrative cases, we tend to demonstrate the phenomenon of decohering worlds with the quantum equivalent of a coin toss, a measurement with two equally probable outcomes. That situation can be illustrated very simply by a symmetrically branching tree. But in general—carefully contrived experiments excepted—different quantum outcomes are not equiprobable. For example, if we make a photon hit an angled sheet of glass, we can make the probability of reflection anything we like just by adjusting the angle, say, 1/7. If, like me, you are a visual thinker, it seems obvious to illustrate this in many-worlds terms by using a tree with branches of proportional width, as in Figure 12-1a.

But this is only a visual metaphor. What are we actually trying to represent by drawing the branches at different widths? Perhaps 12-1b is a better attempt, but it implies that each branch contains multiple distinguishable worlds, which is not the case either. Only two different, distinguishable, worlds have been created by this one quantum event. And in any case, any attempt to generate integer numbers of worlds to get the correct ratios is doomed. If we tilt the glass so as to make the probability of reflection not a simple fraction, but something like $\pi/4$, we will need infinitely large numbers on each side to get exactly the right ratio. Even then we will have problems, because a mathematician will tell you that infinity is just infinity; you cannot have one infinity that is six times as big as another, or indeed any finite ratio.

Infinities always lead to problems. However, let us stomp on one fallacy right away. I have often heard people who should know better say something like this: "If many-worlds implies that an infinity of versions of reality exists, then that must include every conceivable kind of reality, including versions where many-worlds is wrong, or the laws of physics don't work at all." Even the first step in this argument does not hold. Just because a set is infinitely large, it does not need to include everything. For example, the set of all even positive integers {2, 4, 6, 8 . . . .} is infinitely large, but there are many, many categories of things it does not contain. We can instantly see that none of the numbers 7, –4, or 3.14159 are members, for example; nor is the square root of –1. Similarly the mathematics of quantum might imply an infinity of worlds, but that still means only worlds that follow very specific rules.

But coming back to the problem at hand, how can we generate the "correct" answer, which should tell us that we are somehow six times more likely to end up in the right branch than in the left one? When Everett invented the first many-worlds theory back in the 1950s, he simply proposed a concept called "measure." Everett posited that when outcomes diverged (he did not use the term "splitting worlds"), your subjective likelihood of ending up in a particular branch was in proportion to its measure. Many physicists feel that this effectively introduces an extra dimension into the many-worlds representation, justifying the representation in Figure 12-1c, where the measures of the branches are indicated by depth as distinct from width.

This greatly troubles some many-worlds supporters, in particular the notion that measure might imply infiite numbers of worlds. They are concerned about the anti-many worlds argument:

"The only possible reason for accepting the many-worlds formulation, with its absurd extravagance of universes, is its economy of assumptions compared to other explanations of quantum theory. OK, we can interpret Occam's razor to say that we should go primarily for economy of assumptions. Avoiding the need for any new laws of physics is therefore the first priority; ontological economy, postulating the minimum number of worlds, galaxies, universes, or whatever is secondary. So if many-worlds can really explain things with no extra

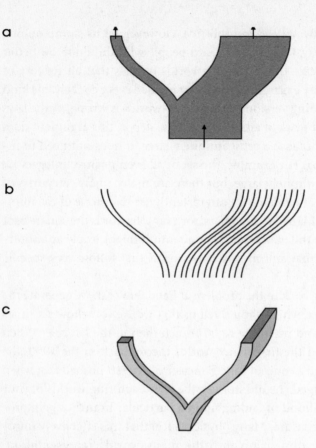

FIGURE 12-1 Branches of unequal probability:
(a) Relative probabilities represented by width of the branches
(b) Realtive probabilities represented by numbers of the branches
(c) Relative probabilities indicated by depth or *measure* of the branches.

physical rules needed, it wins. But if we do, after all, need some new physical assumptions—postulating a kind of extra depth of dimension to reality, for goodness sake!—then the advantage of many-worlds vanishes. In that case it is much more sensible to choose some other interpretation that might need an extra physical postulate but does not also imply an infinity (or at any rate a vast number) of extra universes."

This argument became trickier to refute as it became evident that for Everett's concept to work properly, it may be necessary to make further assumptions about the way measure behaves. We can illustrate with a simple example, as shown in Figure 12-2, where, after inducing one world-branch by tossing a quantum coin, in one branch only we immediately introduce a second branch, with a second quantum coin-toss.

We might naively reason as follows, "There are two distinct branches where the coin came up tails the first time, and only one, in which it came up heads the first time. So at the start of the experiment, it makes sense to bet money the coin will come up tails on the first toss, even if the odds we are offered are less than even—say, if we have to risk a dollar against the chance of winning 70 cents if it is tails." Our intuition rejects the idea that this would be a sensible course of action. But why? To justify turning down the bet, we must make certain mathematical-philosophical assumptions about the way measure works.

All the main defenders of many-worlds have thought long and hard about these problems. The issue has divided them, because although they have answers to offer, in general they are not the same answers. So, let us take a look at these supporters and their camps.

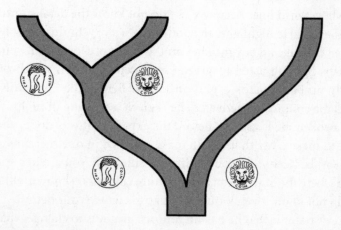

FIGURE 12-2 Consecutive branches.

Lev Vaidman

Lev Vaidman is one of the many counterexamples to the stereotype of theoretical physicists as cold and remote. Very much a family man, he ensures that his sister's violin concerts are advertised in physics department e-mails and occasionally rushes apologetically from a seminar to collect his child from school. Based in Tel Aviv, he recently spent a year as a guest at Oxford University.

Vaidman is a passionate believer in many-worlds and, like David Deutsch, can claim that this way of looking at quantum led him to a technological breakthrough—the Elitzur-Vaidman "bomb tester" was the first zero-interaction quantum measurement device to be constructed. A small, puckish man with a sense of humor, he does not mind telling "Lev" stories that make himself look slightly foolish, if it helps to keep his audience's attention and to get his point across clearly. But he is a theoretician as well as an experimental physicist and, indeed, the author of the authoritative *Stanford Encyclopedia of Philosophy*'s article on many-worlds.[1] His answer to the probability problem is to propose a slight rewording of Everett's original measure postulate as follows:

> The probability of an outcome of a quantum experiment is proportional to the total measure of existence of all worlds with that outcome.

As regards the practical taking of decisions, Vaidman points out that when world lines decohere, we do not know the details until well after the fact. He highlights the point with a story. In this parable, Lev is asked to make an advance bet on the result of a quantum coin toss (perhaps lighting a red or green lamp, depending on which of two equally probable paths a photon takes). Before the apparatus that makes the coin toss is activated, he is given a sleeping draught. When he is awoken, he is asked, "Before the experiment was done, you decided to make a bet that would make you rich in one measure of future worlds. Now you are in a different situation; you are in a specific world where the outcome of the quantum coin toss is known, although you do not know it yet. Would you like to change your bet?"

Lev's point is that he has no rational grounds to change whatever bet he decided to make before the quantum coin toss was done, so there is no practical difference between the classical ignorance interpretation of probability and the quantum all-outcomes-will-actually-happen case. Even in less contrived situations, such as a classical coin toss, it takes time for the different quantum outcomes that are occurring all the time at the microscopic level to be amplified by classical chaos effects to produce sets of worlds sufficiently different that macroscopic events will be different. At a rough guess, the relevant time lag for a difference large enough to make a coin land the other way up might be on the order of 1 minute.

If you bet on the outcome of a classical coin toss and lose, you know that there are worlds containing other versions of you that won—but those other versions had already decohered from your world, about 1 minute earlier. If, on the other hand, you bet in advance on a quantum coin toss that lights a red or green light, by the time you become aware that you have lost, you can assume that you won in worlds that decohered from yours less than a second previously. But your knowledge is always retrospective (because of the finite speed at which neurons fire, and so on), so there is no practical difference between tossing quantum coins and classical ones.

Vaidman could be described as a fundamentalist Everettian, who feels that Everett's original ideas were spot on, and that later concepts—including decoherence, consistent histories, and some of Deutsch's results described below—have been unnecessary to its understanding. He has his own particular take on the question, does measure require large, maybe infinite numbers of each world-line to generate the correct probability ratios. Vaidman has no time for infinities. For him, measure has no more meaning than it is postulated to have. You could perhaps (very loosely) think of it as a kind of tag attached to each world-line with a percentage value written on it, but certainly not in terms of huge stacks of each world-line.

David Deutsch

David Deutsch is to be respected for the courage of his convictions as regards many-worlds. Asking some scientists if they really believe in parallel worlds is a bit like asking a modern theologian if he really believes in miracles; all you discover is that physicists can duck and weave with the best of them. Deutsch does not try to hide behind words or philosophical cop-outs but acknowledges that yes, parallel versions of our world are just as real as our own, including copies in which he himself exists but is doing different things at this moment.

When I first met David Deutsch some years ago, I rather brashly said I wished that he would engage more in the debate between many-worlds and other interpretations. He replied bluntly that he no longer cared to waste time discussing incorrect views. In reality, however, he is a sympathetic man, supportive of his close friend Sarah Lawrence in her work on children's rights, and as willing to talk to students as to those at his own level of knowledge. His manner can be a little disconcerting; he always gives the impression of being highly mentally focused, but not necessarily on his immediate surroundings. Like most of us, his character embodies a certain contradiction; his personal preference for a mildly reclusive existence where he is free to think is often overcome by a genuine desire to help those who want to understand.

Deutsch has made at least three seminal contributions to many-worlds. The first, back in the 1980s, was to use his perspective on many-worlds to formulate a proper architecture for a quantum computer operating on what are now called qubits of information.[2] This led to the foundation of Oxford's Centre for Quantum Computation, where he has remained ever since.

A second contribution was to place the intuitive notion that many-worlds is truly local—that EPR correlations can be explained without involving any kind of faster-than-light influences—on a firm mathematical footing.[3] A third, which we will examine in the final chapter, is a very recent proposal to reconcile quantum theory with the Bekenstein limit, in what he has dubbed "qubit field theory."[4]

Back in the 1980s, Deutsch's original view on the probability question was that it could be satisfied by Everett's notion of measure if we add the postulate that the universe is composed of a continuously infinite-measured set of universes in each of which there is an "I." When a measurement occurs, these universes are partitioned into branches according to the outcome of the measurement.

But recently Deutsch has taken a quite new step. His idea is to start from decision theory, a mathematical way of working out what to do when you are faced with a set of choices. Normally, it is derived from probability theory. But Deutsch has turned the derivation on its head. Starting from very basic assumptions about rational choices (such as that you will be consistent in which results you consider good), he can deduce that you should behave as if you expected outcomes to have relative probabilities in proportion to Everett's original measure concept.

His work has recently—within the past few months as I write—been refined and improved by a young Oxford researcher, David Wallace.[5] I first met Wallace when he was a gifted undergraduate. He has since become one of those polymaths who has mastered all three of the areas: physics, mathematics, and philosophy. He has also found a role working closely with David Deutsch. Wallace and Deutsch have many ideas and attitudes in common; for example, I have heard both independently imply that if we did not live in a multiverse, it would be much more difficult to assign a physical meaning to the concept of probability. A softly spoken man who nevertheless can communicate with sudden and engaging bursts of enthusiasm, Wallace is more willing to attend conferences and engage in roundtable discussions than Deutsch, and the result of their collaboration has been both impressive progress and impressive dissemination of results.

Their joint papers are fiercely mathematical, but Wallace stresses the key result that can be expressed in terms of words: a rational decisionmaker is indifferent as to whether to accept a certain reward or to play a quantum game whose various outcomes equal that reward. We will go a step further and make that statement visual. It means that in a decohering-worlds tree like that shown in Figure 12-2, the cross-sectional area at the top of, say, the left branch is the same as that at the base of the left branch. Taking an extra dummy decision does not really change anything. This generalizes to the proof that the sectional area of any branch of such a tree remains constant as you go up it; breaking it into ever finer twigs never changes its total cross section.

Visually intuitive thinkers might consider this a rather expected result but it is not trivial to obtain mathematically. Figure 12-2 incorporates many simplifications. Branches never really split apart completely, but continue to interact with one another (think of them as connected by a thin skin, like fingers of a webbed hand). The emergence of the basic probability rule of quantum, called the Born rule, is a significant result. What is now called the Deutsch-Wallace program extracts Everett's artificial postulate of measure naturally from the quantum rules, just as it has been found that decoherence does the work once attributed to the artificial concept of splitting worlds, and entanglement does the work once attributed to the artificial concept of quantum collapse.

I do not want to give the falsely rosy impression that all the conceptual problems of many-worlds are solved, however. At least one remains—that the colossal place which is the Hilbert space of the multiverse contains too many possibilities, an embarrassment of riches. Julian Barbour's viewpoint introduces the problem nicely.

Julian Barbour

The loftiest perspective on the multiverse that I know of, in every sense, is offered by Julian Barbour. A tall man with a dignified, patrician English manner, Barbour is representative of a category of scientist that has always existed but is becoming more common in these days of expanded career choice—the researcher who is highly respected by the academic establishment without holding a formal university post.[6]

Barbour's work became known to a wider public a few years ago with the publication of his best-selling book *The End of Time*. I will never forget a public lecture at the London School of Economics that marked the book's launch. Aware that he needed a good gimmick to get the attention of nonspecialists in the audience, he had brought along a bag filled with plastic triangles of various shapes, sizes, and colors, to illustrate his view that the geometry of our universe is best described in terms of triangular distance relationships.

At the appropriate point, he announced, "In this bag I have the basic building blocks of the universe. I think you will be surprised at its contents!" He emptied it dramatically across the stage. However, ahead of the triangles, out bounced a bread roll and several pieces of fruit.

Barbour explained apologetically that he had quite forgotten that he had also placed his lunch in the bag, because it was the only one he had with him. To this day, I have been unable to decide whether this was a supremely clever icebreaker, or merely a supreme example of professorial absent-mindedness. I am only sad that Douglas Adams was not in the room; the incident might have given him new inspiration.

To understand Barbour's timeless perspective on the universe, we must turn again to Hilbert space. We saw earlier how a single point in Hilbert space can represent the state of a system comprising many objects, for example, a single point in a space of about 10^{81} dimensions could represent the state of an entire classical universe. Representing the state of even a single quantum particle exactly, however, requires an infinite number of dimensions, because the particle's position and velocity are describable not by simple numbers but by spread-out probability waves. The Hilbert space which describes the whole quantum multiverse can only be described as mind-bogglingly infinite. Nevertheless, mathematicians can conceive of such a space. You could imagine it as a kind of hazy translucent sphere 10 feet or so across. A single point within that space represents a state of our universe at a particular instant in time.[7]

Some physicists tend to think of this hazy sphere as containing something like a structure of finely branching lines, like those shown in Figure 12-1 which show particular world-histories being traced out in the multiverse. Barbour's insight is that, just as a cine film is in a sense a large collection of still photographs (when they are displayed on a screen at a rate of 25 per second, the sequence gives the illusion of motion), so it is in a sense more accurate to think of Hilbert space as containing a vast collection of snapshots rather than lines corresponding to histories.

But just a moment! *Every* possible state of the universe—every placement of its particles—is represented by some point or other in this hazy sphere. Some of those universes, in fact the vast majority of them, are incredibly unlikely ones. States that belong on what we intuitively think of as probable lines, where time's arrow has triumphed and matter is clumped into stars and planets in an orderly fashion, are a tiny subset of all the points. Tinier subsets still are those patterns containing the illusion of a past history, with features like fossilized dinosaur bones. Why do we find ourselves in such a remarkably special state?

Barbour suggests that we imagine the hazy sphere as being shaded in with a fractal-like pattern of color; densely colored regions represent high-probability states. We can illustrate this in terms of the tick-tack-toe analogy. Rather than an actual game of tick-tack-toe in progress, Barbour sees the multiverse as a sort of computer printout of tick-tack-toe boards containing every possible pattern of X's, O's, and blanks. Boards that embody the history of a legal game, such as shown in the left example below, are much more real (you could think of them as more densely printed) than ghostly boards like that shown in the right example, which of course could not arise in a real game.

So in the real Hilbert space that describes our multiverse, regions that correspond to sensible universe states are much more densely filled in. Universes that encode apparently consistent evidence of a classical history (for example, fossilized dinosaur bones) are in some sense much more probable than random arrangements of matter.

The problem, which Barbour himself highlights, is that it is extremely difficult to see how this probability shading comes about and what it means philosophically. Why do we experience life in a fashion consistent with being parachuted into high-probability regions? His legitimate yet very abstract view of Hilbert space, perhaps the most general perspective that has yet been attempted, highlights how difficult it is to use human intuition to play tick-tack-toe against the gods in such a place.

Murray Gell-Mann and James Hartle

The problem—that Hilbert space describes far too many options—is even worse than we have just admitted. Even in three-dimensional space we know that the same object looked at from different directions can appear quite different—for example, a cylinder can look like a circle end-on, but a rectangle when seen from the side. In infinite-dimensional space the problem is much worse. How to decide which way to draw the axes needed? Why should the directions of the various axes we choose correspond in any way to the directions of our particular three-dimensional space?

The matter gets even more puzzling if we take into account that, according to the mathematics, half the axes represent imaginary numbers—numbers like the square root of minus one. This problem of deciding a preferred set of axes is called the problem of the preferred basis, and physicists wrangle fiercely over whether a unique preferred basis to map Hilbert space to the geometry of our own space-time arises naturally from the mathematics, or must be put in by hand.

Suppose we could peer into Hilbert space with a kind of endoscope or periscope that can be inserted into the hazy sphere at any position and angle. The worlds we could expect to see include not just unlikely versions of our own universe but surreal possibilities like half the square root of minus one times a dead cat plus a live cat. This makes no more sense to a mathematical physicist than it does to a layperson. It is an open question whether, looked at in the right way, such unorthodox viewpoints might even correspond to whole realms of universes that have laws of physics different from our own.

A landmark paper by Murray Gell-Mann and James Hartle builds on an earlier view developed by Robert Griffiths and Roland Omnes, which they called consistent histories.[8] To someone of my views, Griffiths and Omnes's original formulation of consistent histories is a bit like many-worlds with blinkers. We acknowledge that our world is continually influenced by histories beginning to diverge from our own, but, having established mathematically that world lines that start to be macroscopically different from our own have sharply diminishing influence due to decoherence, we simply assume that they vanish. To me, as to other many-worlders, this is a violation of the Copernican principle, an arbitrary assumption that our particular world line is somehow special and unique. It is like saying that because we might never be able to travel to the other planets that we can see through telescopes or to touch and taste things on them the way we can with things here on Earth, we should assume that our Earth is the only real world, at the center of the universe.

Gell-Mann and Hartle, by contrast, are willing to admit the reality of the multiverse—and indeed even the possibility that it contains such exotic things as other realms of world lines. By a clever analysis, they distinguish between what they call weak decoherence and strong decoherence. Weak decoherence creates slightly different world lines that continue to interact (ones where a photon might have gone through a left slit rather than a right, for example). Strong decoherence creates steadily divergent world lines. Their analysis claimed to explain why world lines appear to contain consistent records, that is, patterns that are stable records of events that happened in the past, records that do not change whatever measurements we choose to make. Thus sensible history lines emerge from the jumble of possible states.

Their methodology was challenged by two British theorists, Fay Dowker and Adrian Kent, who reckoned that Gell-Mann and Hartle were in effect assuming much of what they were trying to prove. If you go with Dowker and Kent's viewpoint, Gell-Mann and Hartle's formulation is a bit like the following instructions for getting to Hawaii: "Jump into the Pacific at random. Grab the fluke of the gigantic white whale in front of you that is proceeding in the correct direction."

The point of the metaphor is that within the vast sea of Hilbert space, your chance of finding such a good starting point is much smaller than that of jumping into the Pacific at random and hitting an albino whale. Gell-Mann and Hartle have accepted Dowker and Kent's criticism, but only to a limited extent. The statement in their paper,

[The] persistence of the past is not guaranteed by quantum mechanics alone. Extending a set of histories into the future is a kind of fine graining and this carries the risk of losing decoherence. However, the persistence of the past is critical to the utility of the quasiclassical realm.

now carries the footnote:

Indeed, Dowker and Kent have given examples with special final conditions where a quasiclassical realm cannot be extended at all.

But the question is, are those final conditions really special? Or is it the classical-context cases that are highly special, untypical of general viewpoints in Hilbert space? One defense of Gell-Mann and Hartle's view is a version of what is called the anthropic principle, which is essentially the statement that intelligent beings like ourselves should expect to find themselves in a place capable of supporting the existence of intelligent beings like ourselves. Out of all the possible realms in Hilbert space, it is not surprising that we find ourselves occupying a slice of reality that can support what they call IGUSes, information-gathering and using systems. There might be countless other ways to slice Hilbert space, but obviously we should not expect to see them.

The issue of how to pare down the possibilities of Hilbert space remains controversial.

Conclusion

What is the layperson to make of all this? Are these fine differences of opinion among many-worlders really significant? Certainly there have been what you might call political consequences, because I suspect that if many-worlders had been presenting a more united front, then the many-worlds view would long ago have triumphed.

For what it is worth, my own guess is that the difficulties will be taken care of when it is recognized that the many-worlds view may in some sense require an extra assumption over current physics—but that this is not an insuperable disadvantage. A good defense of many-worlds could be on the lines of Churchill's defense of democracy. At Yalta, Stalin famously asked Churchill how he could possibly be in favor of democracy, given its obvious failings, and the Soviet leader gave a list of disastrous decisions by democratically elected governments. Churchill was at first quite taken aback, but he rallied. "The only thing I can say in defence of democracy, Josef, is this: every other system that has been invented has turned out to be even worse!"

Similarly you could defend the many-worlds view not so much on the grounds of its unique economy as because the alternatives anybody has so far thought of—predestination, cunningly concealed instant links between all parts of the universe, conscious observers with godlike powers to collapse or unmake reality—are all so very much worse. They correspond, at best, to versions of tick-tack-toe that the human mind is ill-suited to play. By contrast, the new many-worlds of Deutsch and his colleagues allows us to play our game with the gods against the backdrop of a universe in which events unfold objectively and locally, in which faster-than-light effects do not operate, and in which quantum probabilities arise naturally, without arbitrariness. The backdrop is a special kind of glass or mirror through which we can see divergent realities clearly enough to use them for measurement and calculation.

Yet as I write, work continues by David Deutsch, David Wallace, Simon Saunders, Harvey Brown and others to see whether even the daunting vastness of Hilbert Space can be conquered and made to yield meaningful probabilities and world lines *without* extra assumptions, just as the more tractable problem of associating probabilities with branches has been solved. That work is still ongoing, but the emerging picture is already a great enough advance on Everett's original concept that it needs a name of its own. Several times I have heard people casually use the phrase "the Oxford interpretation" to describe some aspect of the new work. It is time for the term to be given official status.

CHAPTER 13

THE TERROR OF MANY-WORLDS

Parallel worlds are appealing as an abstract notion, a hypothetical device for making the sums come out right. But if those other worlds are real, then the philosophical consequences are awesome. Every decision that you take must take into account the consequences not just for one you, but for many. For according to the many-worlds hypothesis, the you that exists now will in an instant no longer be a single self but a multitude, each one of them feeling like the sole descendant of the you that exists now. To what extent should you care about the fate of each member of that multitude?

Philosophers have been pondering the puzzles and paradoxes of personal identity—crudely put, what it is that makes you uniquely *you*—since long before the many-worlds hypothesis was invented. They have done so with the help of thought experiments that are distinctly reminiscent of *Star Trek*. Anyone who has done a modern philosophy course might have been challenged with problems like these:

A scientist has developed a machine that can duplicate human beings, complete with their thoughts, memories, and so on. You are told that yesterday, without your knowledge, he duplicated a copy of you. He kept the copy in a lab cell for a few hours, doing IQ tests and so forth, before euthanizing it. How concerned are you to hear this?

The scientist took a copy of you this morning, which he is testing now. He will euthanize it this evening. Of course the copy is protesting that it is the real you. How concerned are you to hear this? Would you be willing to change places with the copy?

The scientist is going to take another copy of you tonight while you sleep. Tomorrow you will wake in your own bed and go about your life as normal, but the copy will awaken in the lab cage and be tested for a few hours before being destroyed like the rest. Knowing this, when you wake tomorrow morning, will you feel scared to open your eyes?

You probably find at least the last of these scenarios alarming. And yet looked at another way, the theoretical possibility that some alien scientist is already making a thousand copies of you every day, and testing them in unpleasant ways before destroying them, can never have the slightest effect on the real you.

It is understandable if your immediate response to the above parable is to resolve to stay well away from deranged alien scientists. But of course many-worlds implies that this kind of duplication of many versions of yourself, who will eventually go on to live out quite different experiences, is a natural process that is unceasing and can never be turned off. Should this be seen as causing problems, or opportunities, for your decision making? Our first example is a tale that has become a classic. It is a challenge that has now been made many times to those who claim to believe the many-worlds hypothesis, and goes something like this:

If you believe in many-worlds, there is an infallible way for you to get very rich. All you need to do is buy a single ticket in a big-money lottery and wire yourself up to a machine that will kill you instantly and painlessly if your ticket does not win. The chance of winning such a lottery is only about 1 in 100 million. But the odds do not matter as long as they are finite. If you believe in many-worlds, then you believe that there is literally an infinite number of versions of yourself in universe-variants that are diverging all the time. After the lottery is run, and the machine has killed you (in an infinite number of worlds) or not killed you (in an infinite number of others), then all the versions of you still alive will be extremely rich.

Of course in a sense, there will now be only 100 millionth as many versions of you as there were before the machine operated. But infinity divided by 100 million, or any other finite number, is still infinity. So in fact there are just as many versions of you as there were before, but now they are all multimillionaires.

As far as I know, no one has yet tried this procedure. But some of the excuses I have heard many-worlders give for declining are disturbingly weak, on the lines of, "I would not like to think of all the versions of my wife and children left poor and grieving in the world-lines where I did not win." This leaves open the question of how to justify declining the option if you have no dependents. Indeed, if you really believe in the logic of quantum suicide, it is arguable that you should seek even more extreme options. Why not wire yourself up to a skullcap containing an EEG that monitors your brain waves to detect whether you are happy and kills you instantly and painlessly at the first hint of pain or sadness? Come to that, why shouldn't we all wear such skullcaps—all 6 billion of us—connected together in a network that painlessly annihilates the whole planet the instant even one person is unhappy? The entire human race would be guaranteed everlasting bliss!

Max Tegmark received a lot of correspondence on the subject of quantum suicide following popular articles in *New Scientist* and *Scientific American,* and has posted the following cautionary note on his Web site.[1]

> I think a successful quantum suicide experiment needs to satisfy three criteria:
>
> 1. The random number generator must be quantum, not classical (deterministic), so that you really enter a superposition of dead and alive.
> 2. It must kill you (or at least make you unconscious) on a timescale shorter than that on which you can become aware of the outcome of the quantum coin toss—otherwise you'll have a very unhappy version of yourself for a second or more who knows he's about to die for sure, and the whole effect gets spoiled.
> 3. It must be virtually certain to really kill you, not just injure you. Most accidents and common causes of death clearly don't satisfy all three.

I do not necessarily agree with him on the first point, because chaos effects very rapidly amplify different quantum outcomes into macroscopic ones. For example, many big-money lotteries use a tumbling cylinder of numbered balls to determine the winning number. Such a machine is a very powerful chaos amplifier, the tiniest difference in, say, the position of an electron on the other side of the world will quickly change the position of the balls. Almost any honest random number generator is rapidly influenced by quantum-level effects.

Tegmark's second point is certainly true, but its implementation is rather trickier. Suppose that your instant suicide machine will not operate until a few minutes after the lottery outcome has been decided. For example, you might have set it up to be triggered by a message from one of those commercial services that send you an e-mail or a text message containing the lottery result. Of course if you see the lottery result before the suicide machine operates, you should be terrified. Presumably you would struggle to escape the machine if you could. But what if you preserve your ignorance by switching off the television, just as if you were trying to avoid seeing a spoiler that would give away the ending of a detective film. How should you feel during the next couple of minutes, knowing that you are now almost certainly going to die, even though many people very similar to you, whose lifelines diverged a few minutes ago, will survive and be happy? I would certainly be terrified—I would want to be unconscious under deep anesthesia for this period.

Tegmark's third point I unreservedly agree with. The chances of winning a big-money lottery are very tiny, on the order of 1 in 100 million to 1 in a billion. That is much smaller than the per-flight risk of being killed in an airplane crash, or the per-lifetime risk of being hit on the head by a falling space rock. In fact, when you next buy a lottery ticket (if you are in the habit of doing so), you might like to reflect that even without going to the trouble of constructing a diabolical suicide machine, you are a lot more likely to be killed in a bizarre accident before the lottery is run than you are to win it. If your lottery machine has, say, a 1 percent chance of malfunctioning and leaving you injured or brain-damaged rather than dead, then your rational expectation is a million to one that you will emerge from the experiment poor and crippled rather than intact and rich. To make it a hundred times more likely that you would survive rich than survive poor and crippled, the mechanism would have to have less than 1 chance in 10^{10} of failing during operation. I doubt that any comparable machine constructed by humans has achieved that level of reliability, much less a novel design that has not been tested in full operation.

There is a much more worrying corollary to this lottery story, which was articulated by the philosopher David Lewis. He pondered the fact that in a quantum multiverse, every possible cause of death is just a variant of this style of Russian roulette.

For example, suppose you die of being run over by a truck when you cross the road in a hurry without looking properly. A very tiny change in events might have spared your life. For example, the human retina is potentially sensitive to the impact of individual photons, though the neural processing circuits in your optical nerve usually screen out such tiny fluctuations. But the impact of a single extra photon might have tipped those neural circuits into warning your brain of a fast-moving object in your peripheral vision and saved your life. There will be countless parallel worlds where that occurred.

Even once you are physically in the path of the truck, your death is far from certain. The trajectories of the air molecules around you might add up so as to cause them to give you a sideways push just before the truck hit you, in a scaled-up version of Brownian motion, reducing the impact to a survivable level. Of course that is very unlikely; Brownian motion normally affects only tiny objects in this way. The odds against it might be on the order of 1 in 10^{100}. But it is physically possible right up to the last instant before the truck hits you, and that still leaves an infinity of survivors. Even after the truck has hit you, the molecules in your body might bounce around in such a way that your tissues are not destroyed, all accelerating in perfect synchrony. And so on. David Lewis reasoned that there would always be surviving variants of you in some of the subsequent physically possible histories, and feared the implications.

Previous thinkers who had the same idea (certainly Huw Price, anecdotally, many others) welcomed it as a delightful discovery. We are immortal, our consciousness can never be extinguished, rejoice! But we can all remember childhood fairy stories where people are granted magical wishes by some genie or fairy godmother, make ill-thought-out choices, and regret them. An error that appears in many such stories is to wish for immortality, but forget to wish for perfect health and youth, so that you get ever older, iller, and more infirm without the ultimate relief of death. David Lewis realized that even if a truck cannot kill you, it can still maim. In fact a freak bouncing of molecules just sufficient to spare your life (but leaving you horribly crippled) is vastly more likely, relatively speaking, than one that leaves you altogether unscathed. He feared that we were all caught in the horrible trap of the fairy story just described, and pointed out that though we might devoutly wish we could die, the rules of the universe do not follow our wishes. Maybe we are all doomed to live forever.

In a paper "How Many Lives Has Schrödinger's Cat?" delivered in Canberra in June 2001, Lewis made his views clear. His lecture ended with these chilling words[2]: "What you should predominantly expect, if the no-collapse hypothesis is true, is cumulative deterioration that stops just short of death. The fate that awaits us all is the fate of the Struldbruggs [the immortals in Jonathan Swift's *Gulliver's Travels*]....[3] How many lives has Schrödinger's cat? If there are no collapses, life everlasting. But soon, life is not at all worth living. That, and not the risk of sudden death, is the real reason to pity Schrödinger's kitty."

Although his words are light, I am told by those who worked with him that he was terrified by this hypothesis.[4] By a cruel coincidence, he died suddenly and unexpectedly from diabetes within weeks of giving that lecture—at least in our version of reality. His paper is about to be published posthumously as I write. He must have died a badly frightened man, and the psychological impact on his colleagues was considerable.

Should we really fear becoming Struldbruggs? A year ago, I was at a seminar where David Deutsch was asked whether he feared this scenario. His answer was that he did not fear world lines in which he might enjoy a very extended life, because in the vast majority of such instances, this would come about due to advances in science and medicine in which he would be voluntarily enjoying a reasonably healthy existence. To an extent I can see his point. After all, a world in which remarkably unlikely medical breakthroughs have occurred is far less improbable than one where remarkable second-by-second violations of the usual statistics of Brownian motion conspire to keep your brain indefinitely alive in a body that has effectively ceased to function. But I am not entirely reassured. Even in our presumably high-probability world line, large numbers of people are already being kept alive long after the point where their quality of life has become negative. In any case, the putative coming into existence of large measures of worlds where I am long-lived and happy does not comfort me about what I will inevitably experience when I am finally hit by a truck, or suffer some comparable accident normally considered life terminating.

Max Tegmark does not fear the cannot-die scenario for the more comforting reason that the fading of consciousness is a continuous process. Although I cannot experience a world line in which I am altogether absent, I can enter one in which my speed of thought is diminishing, my memory and other faculties fading, as happens gradually in old age, and rapidly but not instantly if you become unconscious from more immediate causes. He is confident that even if he cannot die all at once, he can fade gently away.

David Wallace puts a similar argument in a slightly different way, invoking extension in space rather than extension in time—our consciousness is not located at one unique point in the brain, but is presumably a kind of emergent or holistic property of a sufficiently large group of neurons.[5] Thus the left half of my brain, containing a certain degree of consciousness, can enter a world line where the right half has just been crushed by a truck. A group of 1,000 neurons in my hippocampus can enter a world line where the rest of my brain has been destroyed, and so on. Again the prediction is that our consciousness might not be able to go out like a light, but it can dwindle exponentially until it is, for all practical purposes, gone.

Just in case you are now feeling too comfortable, there is a second quite different, but almost equally nightmarish, implication of many-worlds. You recall that part of the solution to the problem of picking out sensible worlds from the infinite choice that the equations of Hilbert space describe is that it is only those world lines where the laws of physics continue to work sensibly that can contain IGUSes, in other words, conscious entities like ourselves. It is an understandable prejudice that these are the only lines that are worth thinking about and that alternative snapshots of reality implicit in the equations can simply be ignored.

We have a strong subjective prejudice that the number of versions of reality is in some sense increasing. As time passes, the multiverse seems to generate more and more realities incorporating the you of the present moment, tracing out different future histories. But this dear reader, is not the whole story. It is equally conceivable that corresponding to the present you, there will develop not only versions of you that will continue to exist long term in diverging but sensible world lines, but other you's that are doomed to rapid extinction in lines where the laws of physics are ceasing to operate consistently. As Michael Lockwood, Simon Saunders, and others have pointed out, evolution is driven by the ability of IGUS-like patterns to preserve and reproduce themselves in worlds that continue to follow sensible rules, and so evolution inevitably designs our brains to cope with those lines.

But what if there are discards, patterns that remain self-aware for at least a little while in a universe that is ceasing to obey the familiar rules? Perhaps in each second of your life, for every you that continues to enjoy a familiar existence, there are created an infinite number of failing versions who have time to wonder what is going wrong before their existence fades out, in something like the manner described by Thomas Disch in the classic *Echo Round His Bones.*

The surviving yous would never become aware of this process, of course, just as you would never become aware of the activities of the hypothetical alien we posited at the start of the chapter who persistently takes copies of you and subjects them to different fates. And do not be falsely reassured merely because the version of you now reading this book has been a winner for many years, always one of the versions that stayed in a sensible reality. Consider fish like sturgeon, which lay hundreds of thousands of eggs which develop into free-swimming larvae; on average, only two—one male, one female—will grow up and breed successfully. (Biologists know this because if the net population growth per generation were even a fraction of 1 percent, the ever-growing total mass over the millions of generations of fish that have occurred would soon vastly exceed the total amount of organic matter available on Earth.) Imagine how confident each larva could feel, knowing that not one of its ancestors has ever been eaten before reaching breeding age, in an unbroken line of succession stretching back millions of years. Yet in a sea full of predators, the life expectancy of most of the larvae is measured in hours rather than days. We could be in an even more extreme version of their predicament.

Do such cast-off versions of you really exist? At present, I don't think anybody can meaningfully answer that question. Sweet dreams. . . .

Now to a more positive prospect. There is a classic problem involving personal identity and probability that appeals to many-worlders for several reasons, but especially because it might be more straightforward to solve in a many-worlds context than in a classical single reality. It is nowadays called the Sleeping Beauty problem, although it was first written up in 1997 as the Paradox of the Absent-minded Driver, and an oral version might be older than that. [6]

The story is that you volunteer to be a human guinea pig for an experiment with the following procedure. You will be given a drug that will put you to sleep for a short period. While you are asleep, the experimenter will toss a coin. If it comes up tails, he will awaken you, and that will be the end of the experiment; you will go on your way. But if it comes up heads, he will awaken you and then ask you to swallow a second pill. This one will put you to sleep a second time and also erase your short-term memory so that you have no memory of the brief period of awakening. (There are real medicines that work very like this, such as the infamous "date-rape" drug Rohypnol.) After the experimenter wakes you from this second period of sleep—of course, you will have no way to know it is your second awakening—the experiment will end and you will go on your way. Figure 13-1 shows the two ways that events can proceed.

The experimenter explains that on every awakening you will be asked a simple logic question to see whether your thought processes are working clearly. This all seems harmless enough, so you swallow the first pill and lie down on the scientist's lab couch. In due course you awaken.

"I would like to ask you the following question," says the scientist. "What is the probability that the coin fell heads up?"

FIGURE 13-1 Quantum Sleeping Beauty.

You ponder. Surely the answer must be simply one-half, assuming it was a fair coin. But then a curious point occurs to you. If the coin fell heads up, there are two occasions on which the scientist will ask you this question. But if the coin fell tails up, there will only be one such occasion. There is therefore a good case that the correct answer is two-thirds!

The problem can be made more dramatic if we raise the stakes a little. Suppose that instead of tossing a coin, the scientist spins a roulette wheel with 100 numbers on it. He tells you that if it comes up one particular number, he will wake you and put you back to sleep 10,000 times before ultimately killing you! However, if it comes up any other number, he will let you go on your way unharmed after waking the first time.

You are forced or tricked into taking the first sleeping pill. You awaken. How scared should you feel? By one reckoning, the chances are 99 percent that the roulette wheel spared you and you will shortly be allowed to walk away. But by another reckoning, if you draw an outcome tree like that in Figure 13-1, there are 10,000 awakening-instances on the fatal branch of the tree, and only 99 on the branches in which you survive, so you are probably doomed.

If you tend toward the optimistic point of view—on awakening in this second experiment, you would feel 99 percent confident of survival—let me introduce a slight variation that does not really change the odds at all. The room in which the experiment is done contains an independent witness who observes every awakening of every subject the scientist does this experiment on. (He repeats the experiment with hundreds of subjects, most of whom of course survive.) As you awaken, you see the witness observing you with an enigmatic expression before getting up from her chair and leaving, because although she knows the roulette-wheel outcome, she is not allowed to give you any clue. Your blood chills as you realize that she has gotten up from her chair like this on thousands of occasions, and on 99 percent of those occasions the subject before her has been doomed. It seems that whereas before going to sleep, you were 99 percent confident of survival, on awakening you should feel very afraid. . . .

The Sleeping Beauty problem, as it is called, has no agreed-upon answer. But Lev Vaidman has written a paper in which he claims that he and Simon Saunders, two many-worlders, have a straightforward answer to the first case if the coin is replaced by a quantum-random device, such as a photon which can be absorbed or reflected.[7] Because then the ignorance interpretation of probability does not apply; both world lines have equal measures of existence by Everett's rules and hence so does each of the three episodes of awakening. When you awaken, what mathematicians call your rational expectation that the coin fell heads up should be two-thirds. By consistency, the answer should be the same even if a classical randomizing device such as a coin was in fact used. Similarly, in the second case, when you awaken, your expectation that you will survive should be only 1 percent.

This answer to the Sleeping Beauty problem remains controversial. But the tale is at least a charming classical introduction to the kind of philosophical conundrums that arise in many-worlds reasoning about the self.

⊗⊗⊗

At the moment the philosophy of many-worlds clearly contains more questions than answers. But in terms of practical everyday choices, where does this leave the reader? The best I can do is to quote the opinions of those most knowledgeable in the field. A year or so ago, sitting next to David Deutsch at a dinner, I had the chance to ask him,

"Is there *any* decision that you would take differently on account of believing that you are in a many-worlds universe, rather than in a classical one?"

Rather typically, he smiled and answered, "Yes. I would answer the question, 'Do you believe that you are in a many-worlds universe?' differently."

But on being pressed, he gave his more serious answer, which boils down to No. He believes that it would be crazy to behave in any other way than in proportion to the measure of existence of the possible worlds consequent on your actions. In every practical life decision, including those involving gambling using either classical or quantum random number generators and those that involve risk—the possible termination of his own existence—he would make exactly the same choices in a quantum multiverse as in a classical universe. Most quantum thinkers whose opinions I respect agree with him.

But there are dissenters. Some physicists are more equivocal, commenting that in their old age, with no remaining dependents and relatively little life expectancy at stake, they might be tempted to some form of the quantum-roulette gamble. To me this is merely a version of what I call the "Krakatoa argument." If you know that the volcano on an island is about to blow shortly, and your available funds are only half the amount needed to buy a place on the last boat out, then it is perfectly reasonable to go into the casino and bet all your money on the red—even if the odds are slightly poorer than even. In the same spirit, if you reach a point of old age and infirmity where only immensely costly medical and nursing care would improve your quality of life to a level where it was worthwhile to continue, the time might come to attempt the quantum-roulette gamble. Of course you could also argue that by then the roulette gamble is a sensible choice even if you believe only in a single world.

Another advantage of postponing your decision for as long as possible is that, by then, physicists and philosophers may have revised their advice. Many-worlds is not yet proven. And there is the possibility that we live in a multiverse of finitely many worlds, a possibility we will consider in the last chapter. That makes a fundamental difference to quantum Russian roulette and similar games. Infinity divided by 100 million is exactly the same infinity as before. But a merely very large number divided by 100 million is that many times smaller; for example, 10^{100} divided by 100 million shrinks to 10^{92}, killing the overwhelming majority of potential future yous. Personally, I will be sticking with Deutsch's advice.

CHAPTER 14

THE CLASSICAL WARRIOR

Roger Penrose

What if quantum theory is, after all, incomplete? What if there is some as-yet-undiscovered physical mechanism that can bring about quantum collapse, and by implication undermines the case for many-worlds? This is now a minority view, but it is a possibility that some physicists still take seriously.

The first reasonably watertight specification for such a collapse principle was formulated in the 1980s by three Italian physicists—Ghiradi, Rimini, and Weber—following a program suggested by Philip Pearle. In their honor, specific physical collapse mechanisms postulated ever since tend to be referred to as GRW-based mechanisms. Their basic point was very simple. Systems in which quantum behavior had at that point been observed involved very small numbers of fundamental particles, most typically one (as in the two-slit experiments that had been performed by that point), or at most just a few hundred. Suppose that there is some mechanism that collapses the quantum wave function at random intervals—so that, for example, an object described by a spread-out wave function suddenly pops back into existence in a single well-defined place—but the mechanism works in such a way that individual particles are collapsed only at very long intervals, yet systems containing huge numbers of such particles are collapsed at very short ones.

An appropriate collapse probability might be on the order of 10^{-16} per particle per second. Obviously, the chance of a particle collapsing into a single position in the tiny fraction of a second during which it flies through a two-slit experiment (or any other experiment done on a human timescale) is utterly negligible, so wavelike behavior is observed. On the other hand, anything large enough to be considered as a classical measuring device—an observer, be it a cat, a human, or a laboratory instrument—contains something on the order of 10^{24} to 10^{29} particles. Accordingly, such a system would be expected to collapse to a single location and state in a tiny fraction of a microsecond. Since GRW put forward their program, several people have suggested specific candidates for such a collapse mechanism.

The most prominent of them is Professor Sir Roger Penrose, and his ideas deserve special consideration because he has done considerable work devising experiments which could actively verify them. Now in his mid-70s, Penrose is one of those scientists who has remained remarkably undiminished by age; in both mental and physical agility, he could easily pass for a man in his 50s. In many ways he is the last and most impressive representative of the old school of quantum thought. He epitomizes it especially well because on the one hand, he is seeking a very specific and reductionist physical mechanism to explain quantum collapse; yet on the other, he assigns at least as much importance to more philosophical issues, and specifically to the possibility of a link between quantum and the nature of human consciousness. To me, and perhaps to others who embrace the new school of thought, this is both slightly paradoxical and eerily reminiscent of the attitudes of such past figures as von Neumann and Bohm.

Penrose is now retired from his distinguished post as Rouse Ball Professor of Applied Mathematics at Oxford University. The achievements that first raised him to worldwide prominence date to the 1970s. Their common link is geometry, because his ability at mathematics is combined with extraordinary visual insight. At the recreational level, he has invented paradoxical shapes of the type made famous by Escher. His discovery of ways to tile a plane in a pattern that is nonrecurring, infinitely varied, and not predictable by any computer algorithm is a beautiful illustration of a deep problem in mathematics. In physics, his work on matrices called twistors has suggested ways in which the warped fabric of space-time described by general relativity might be reconciled with quantum theory. And I am just old enough to remember the furor when he and Stephen Hawking first revealed their work on the nature of black holes.

These achievements were a third of a century ago. Yet Penrose today is more famous even than he was then. I had a recent reminder of this when he was due to give a talk in a lecture hall that is usually amply large enough for the seminars held there. I turned up well in advance to find it full to bursting. Not only was every seat occupied, but the entrances, stairways, aisles, and even the platform at the front were packed solid with young students willing to stand or crouch in uncomfortable positions for the privilege of hearing him speak. It would have been quite impossible for Penrose himself to get into the room, and we all had to trek across town to a much larger lecture theatre before the talk could go ahead.

Most of that audience had not been born when Penrose made the discoveries for which history will remember him. The work that has brought him back into the public eye is on a significantly different topic. Like many physicists, in his later years he has become increasingly interested in more philosophical issues, questions that are hard not merely to answer but even to formulate. This new focus has led him to propose not one, but two, controversial hypotheses concerning quantum theory, which have led him into significant conflict with his colleagues even as they have raised his profile with the wider public. There is now a significant rift, which has cultural as well as ideological dimensions, between him and the newer generation of physicists. His views on foundational issues, the bedrock on which physics is grounded, differ profoundly from those of many younger scientists. I once dared to bring up David Deutsch's name in a discussion on the philosophy of mathematics, to be told sharply: "David and I seem to disagree on just about every conceivable point."

But Penrose's enduring creativity and mental sharpness are not in doubt. The willingness to formulate new hypotheses, to challenge established wisdom wherever puzzles remain, is essential to scientific progress. With respect to quantum physics, Penrose is in the rare position of having been active through both the major epochs in which quantum interpretations were generated. Earlier in his career he knew David Bohm when they both worked at Birkbeck College in London, yet he has remained active and innovative right through to the present day. His views certainly deserve a hearing, and in this chapter we will examine both of his major hypotheses with respect to quantum.

Collapse by Gravity

Penrose has gone looking for a plausible mechanism that might cause collapse along the lines suggested by GRW, and found an answer suggested by a well-known dichotomy between quantum theory and general relativity. It is known that if general relativistic effects (broadly speaking, gravity) were subject to quantum fluctuations in the way that other fields and energies are, mathematical infinities would arise. In physical terms, the structure of space-time would be violently unstable. A quantum fluctuation in a tiny region of space-time would very rapidly grow, perhaps spawning exotic entities, such as black holes or wormholes, at a colossal rate. We do not observe anything like this, so we know that at least some correction is required to current theory. Penrose has tried to fix two problems with one solution by suggesting that such uncertainties in gravitational field energy tend to cancel themselves out, producing the process we call quantum collapse as a side effect.

Gravitational attraction, in the modern Einsteinian picture, is caused by a warping of space-time. In a well-known analogy, this can be crudely visualized as like the bending of a rubber sheet when a heavy ball is placed on it. The heavy ball makes a dimple, and smaller balls placed on the sheet tend to roll down into the dimple just as small objects tend to fall toward the surface of a planet. Even though we normally think of gravitational pull as associated with large objects, such as Earth or another planet, in fact all matter produces gravitational effects. Even two atomic nuclei attract one another gravitationally, and therefore produce tiny dimples in space-time.

If you reject the guide-wave hypothesis (as most modern physicists, including Penrose, do), then an object whose position has acquired a wavelike uncertainty really can be thought of as being in two or more places at once. But what about its associated gravitational field? If the object in Figure 14-1 could be in either of two positions, does it curve space-time as in Figure 14-1a, Figure 14-1b, Figure 14-1c, or what?

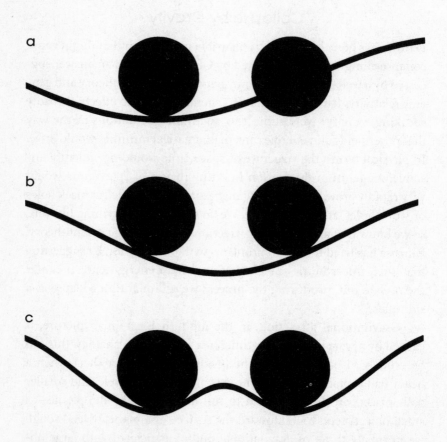

FIGURE 14-1 (a) Space curves as if only one version of the ball (say, the left one) has real mass.
(b) Space curves as if the mass of the ball is at the midpoint of its two possible positions.
(c) Space curves as if there are two versions of the ball, each with real mass.

Penrose's hypothesis can be expressed in terms of the picture of Figure 14-1c, where space-time starts to be deformed as if by two different objects. Like an elastic band being stretched, this kind of double dimple stores energy. Penrose suggests that the fabric of space-time resists this kind of thing, and that the higher the energy stored in the double dimple, the more likely quantum collapse is to pop the object into a well-defined location.

How great an uncertainty in the gravitational field is needed to cause quantum collapse? Here Penrose resorts to an admitted guess. He speculates that the collapse time, T, in an isolated system (that is, one that is not currently interacting with or being observed by a larger system) is of the order $h/2\pi E$, where h is Planck's constant, the tiny quantity we met in Chapter 2, and E is the energy that would be released by allowing the two versions of the object to fall to their common center of gravity.

For everyday masses—billiard balls, say—the expected collapse time, T, would of course be very short indeed. However, it becomes more significant for tiny objects, and Penrose has calculated the following approximate collapse times:

Beryllium ion ~ 100 years
Water drop 2-μm diameter (just visible in microscope) ~
    1/20 second
Cat ~ 10^{-37} second

You may find it interesting to compare these with the collapse-by-decoherence times given in Table 7-1.

There are hand-waving elements to Penrose's argument, but one very important thing sets him ahead of others who have proposed GRW-type collapse mechanisms. He has been brave enough to propose a method of testing his theory, and put considerable effort into refining it toward practicality. By chance I was privileged to hear his very first public description of the suggested experiment in a lecture to Oxford students, the day before he gave it a more formal presentation at Imperial College in London.

The proposed apparatus is our old friend the Mach-Zender interferometer, which we met being used as a bomb detector in Chapter 10, illustrated in Figure 10-1. Wavelike behavior in this device ensures that the photons all arrive at detector E. Wavelike behavior of course requires that we do not make any measurement of which of the two possible routes the photon takes through the main part of the apparatus.

Suppose we replace the fixed mirror at B with one that is free to move? It will now recoil slowly if the photon hits it, but the effect is so tiny that under normal circumstances it will be swamped by other uncertainties in the mirror's position and motion, so no interference-destroying "measurement" will take place.

But what if we make the distances BD and AC very large—on the order of thousands of kilometers? Then the mirror will have time to move a significant distance while the photon is still in flight. In fact, if Penrose's theory is correct, the uncertainty in the mirror's position caused by the lack of definiteness as to whether the photon went via mirror B or by the alternative route becomes so large that gravitationally induced collapse occurs. The mirror "pops" into one position or the other. At that moment the photon gets localized on one route or the other. When it arrives at the detectors, it will behave in a particle-like way, with an equal chance of being detected at E or at F. This differs from the predictions of orthodox quantum theory.

This would be a very difficult experiment to do. The only practicable way to get the long photon path lengths required would be to mount the experiment aboard a pair of satellites, which would of course be very expensive. Penrose speculated on ways to get around this problem. One possibility would be to use an X-ray or gamma-ray photon, whose energy and momentum is much higher than a visible photon. Unfortunately, it is also much more difficult to generate and handle such photons.

For several years, Penrose (whose retirement has been more nominal that actual) worked with Oxford graduate student William Marshall and others on ways to make the experiment practicable. One possibility Marshall told me they were exploring involved bouncing the photon repeatedly between two closely spaced mirrors on each leg of the apparatus. Using mirrors tuned to the relevant wavelength to create what is called a high-finesse cavity, the photon could be made to bounce millions of times before continuing on its way. This would have two benefits. The first is to increase the delay time before the photon is measured to a reasonable value while keeping the apparatus quite compact. The second is that if the mirror that is allowed to move is one of the cavity mirrors, it will get kicked not once by the photon, but a huge number of times. (Think of a tennis ball bouncing rapidly between your racket and the ground when you hold your racket close to the ground.) The repeated photon bounces have a much bigger effect on the position of the mirror than a single reflection would do. If the mirror is mounted on a flexible support, a silicon cantilever, and made to vibrate to and fro to start with, the effect of the photon could in theory cause it to end up in a significantly different position to that it would otherwise have occupied, at the opposite end of its swing. By Penrose's argument, the mirror will spontaneously collapse itself into one of those two positions, thereby determining definitely which cavity the photon is in and abolishing interference effects.

In 2002, Penrose and Marshall published a paper describing this more sophisticated version of the experiment.[1] However, the authors' own calculations show that using off-the-shelf technology, it would be about 100,000 times less sensitive than required to prove Penrose's hypothesis. Definitive results will not be coming anytime soon.

In the absence of experimental proof, what is the current establishment's verdict on Penrose's gravitational collapse? It has to be said that it is fairly dismissive. Stephen Hawking probably speaks for many when he says that decoherence explains collapse so well, without needing to invoke any new physics, that it has simply become superfluous to look for alternative mechanisms. My own feeling is that this might be a little harsh. At the very least, Penrose's highlighting of the fact that different quantum outcomes can rapidly lead to presumed interference between outcome worlds where the very shape of space-time is significantly different is worthy of further pondering, and experimental investigation if possible.

The bottom line, however, is that Penrose's collapse mechanism does not resolve what we have identified as the one true dilemma of quantum theory: the nonlocal nature of collapse. Penrose agrees that if we were to entangle two systems and move them a light-year apart, collapsing one of the systems—by observation or by gravity—would instantaneously change the expected results of a measurement on the second system. It would seem that you could send a "forbidden" faster-than-light message in this way; by fiddling with the mass that triggers the quantum collapse, you could select which of the two distant detectors the photon would arrive in, right up to the last moment.

This does not necessarily cause paradox, however. The backward-in-time signaling we discussed in Chapter 4 depended on the fact that we had two pairs of faster-than-light signalers in two different frames of reference, aboard trains moving in different directions. There are still a few physicists who hope that, special relativity notwithstanding, the universe will turn out to have one preferred stationary frame of reference after all, violating the principle called Lorentz invariance—that the laws of physics look the same to all particles, whatever their velocity. If there were such a unique frame of reference, then faster-than-light information transmission with respect to that frame only would not equate to backward-in-time signaling, and would not cause paradoxes. One such model was formulated in 1949 by Howard Robertson of the California Institute of Technology, and developed further in the 1970s by Reza Mansouri and Roman Sexl of the University of Vienna. As recently as 1998, a set of Lorentz-violating interactions was postulated by Sidney Coleman and Sheldon Glashow at Harvard University. But no evidence for Lorentz violation has ever been discovered, and conventional relativity remains, to put it mildly, the overwhelmingly more accepted paradigm.

Penrose's genius notwithstanding, his fondness for his gravitational-collapse hypothesis might, at the end of the day, reflect the fact that most of his life was lived before the experiments of Aspect and others, which have unequivocally proved that nonlocality is real. Before nonlocality was proven, finding a plausible collapse-causing mechanism was perhaps the most urgent problem of quantum theory. But now nonlocality must be faced, and no local collapse mechanism, however cleverly devised, can appease its dragons.

Collapse in Mind

Penrose's second hypothesis about quantum collapse is very much more speculative than the first. It is ironic that thanks to his popular books, in particular *The Emperor's New Mind*,[1] it is by far the best known of his ideas to the general public. In an earlier chapter, we mentioned the dubious hypothesis that conscious observers play a special role in the establishment of reality. Later, we discussed the likely manufacture of quantum computers in the near future. Somewhere between these two poles of wild and solid speculation comes Penrose's notion that the human mind might itself be a quantum computer. His declared motive is to explain how our minds can have certain capabilities that he claims would be impossible for any computer operating according to the principles of classical physics.

The vast majority of scientists today accept that the human brain is a form of computer. Of course it differs from the one on your desktop in many ways. The most striking is that your computer has a single processing unit that is doing just one thing at any one time, whereas your brain consists of some 100 billion neurons all operating at once, each acting like an independent computer that reads electrical signals from up to 10,000 other neurons it is hooked up to on the input side, and then broadcasts its own signal to a different batch of neurons on the output side. Some neurons connect to locations outside your brain, for example, receiving signals from the retinal cells in your eye, or telling muscles to contract. Thus your brain is also able to interact with the external world.

But does the power of 100 billion processors make your brain fundamentally different from a desktop computer? The answer turns out to be no. One of the foundations of modern computer theory, invented by the British mathematician Turing during the Second World War, is that any computer that operates according to the laws of classical physics, massively parallel or otherwise, can be exactly simulated by a very basic computer capable of executing just one simple instruction at a time, provided you give it enough time and enough memory with which to work. Such a machine can be—and has been—built from a simple construction toy like Lego, yet it can in principle exactly simulate the workings of any computer based on traditional physics, from your desktop PC to an organic brain.

If you accept that your brain works by classical physics, then the only difference between it and your desktop PC is scale and programming. We already know how to program an electronic computer to simulate the workings of a small group of neurons performing a simple task, but to simulate the human brain a computer would require at least 10^{17} binary digits of memory. The computer on your desktop probably has a memory size on the order of 10^9 binary digits, insufficient to simulate the brain of an insect, even if we knew how to program it appropriately. The task of replicating the human brain is still—thankfully, for moral reasons—way beyond us.

But Penrose feels strongly that the difference between the human brain and a computer is more than mere scale and programming. He believes that human intuition, or more precisely what he regards as the ability of our minds to transcend algorithmic reasoning (that is, step-by-step reasoning using a fixed set of rules) proves that there must be some beyond-Turing-machine aspect to our minds. He describes in particular the "aha" moment when we have been worrying at some problem in an unimaginative step-by-step way without making any progress, then suddenly a lateral-thinking method of going forward, by seeing things in a new way, seems to pop into our heads without warning. To him, this seemingly instant condensation of nebulous thoughts into a coherent solution is strongly reminiscent of quantum collapse.

Penrose highlights one example that he feels proves his point. It arises from an attempt 100 years ago by the remarkable mathematician Hilbert (whose Hilbert space and Hilbert hotel we have already met) to formulate a mathematical language with a comprehensive set of axioms and rules that, within the context of a given mathematical system, will allow any proposition—that is, any grammatically meaningful statement—to be explicitly demonstrated to be either true or false. Because such a language contains a finite number of symbols, there is a finite number of statements of given maximum length that can be made. Because there are also a finite number of rules for manipulating the symbols, an appropriately programmed computer could easily fiddle about with the axioms to deduce further true statements from them—it becomes an entirely mechanical process. Conversely, the computer could fiddle about with any arbitrary statement it was given until it was either reduced to some combination of the axioms, or shown to contradict one or more of them.

However, a perfect system of this kind, in which every proposition can be proved true or false by applying such a sequence of steps, turned out to be very elusive. Finally a mathematician called Gödel discovered a remarkable thing. Any such system must necessarily contain some statements that are in fact true, but that can never be proved within the rules of the system. The essence of his proof was to list all the possible propositions that can be made, and all the possible correctly formulated proofs, in a kind of alphabetical order, demonstrating that some of the proofs we might expect to find will inevitably be missing.

Penrose claims that a Turing-machine type of artificial intelligence would find it impossible to understand how such a Gödel-undecidable statement might in fact be true, despite lacking a formal proof in terms of the rules. Other people, myself included, cannot really see what he is driving at. What do we mean by "true"? Each of us has an intuitive definition, arising from the experience and mental development of a lifetime. When we are given a mathematical rule system of the kind described above, we can temporarily accept a redefinition of truth as "provable by manipulating the symbols according to certain specified rules." But of course we have not really forgotten our broader notion of truth, and when we find the rule system inadequate, we appeal to that broader intuition.

Penrose might well be right in his feeling that a human "aha" moment of intuition represents a collapse of multiple tentative threads of thought into a single successful perception. And this is certainly analogous to what happens when the parallel "thought processes" of a quantum computer collapse into a single outcome at the end of the computation. But of course there is no need to invoke quantum to explain why the brain is capable of massively parallel processing. We noted at the start of this section that the brain has a hundred billion neurons at its disposal. Of course our brains use parallel processing, and no doubt our apparent thread of consciousness is a retrospectively constructed story in which the work of unsuccessful subnetworks is jettisoned, and we remember only the reports of the successful networks—this is rapidly becoming the consensus view among neuropsychologists.

Gödel's theorem is a fascinating mathematical result with real practical implications, in particular that there might be many reasonable-sounding problems that a conventionally programmed computer would require an infinite time to solve.[2] Penrose's appeal to quantum seems to be based on the hope that some kind of ultra clever quantum collapse might give our minds flashes of intuition about those beyond-infinity solutions. But when we come to study real quantum computers, we see that they outperform classical ones only quantitatively rather than qualitatively. The advantage can potentially be impressive, but it is never infinite. Penrose himself seems to acknowledge that to return information about results that could not be found in finite time by a Turing-machine computer, quantum collapse would have to possess properties additional to and even weirder than those it is already known to have.

What has made Penrose's quantum consciousness so popular with the public, and inspired him to work on it for so long? Probably it is the enduring longing to believe that the physical basis, as distinct from the mere software, of human beings in some way transcends our mundane material world. Once, all living things were assumed to be endowed with some special vital force. As experiments probed first animal and then human cadavers, the body was seen to be mere ingenious machinery. The physical mystery retreated toward a last hideout somewhere in the brain. Around the time I was born, some doctors were still trying to weigh bodies at the point of death, attempting to detect a small reduction in the weight as the soul departed. (They did find a tiny reduction following the last breath. I suspect it was the buoyancy of the body-temperature air in the lungs, lifting the body like a hot-air balloon with a force of a fraction of a gram.) The human mind is a wonderful thing, but it needs no unique physics to explain it.

CHAPTER 15

THE NEW AGE WARRIOR
Anton Zeilinger

There is a hoary old joke: What do you call a physicist who works on quantum theory? Why, a quantum mechanic, of course! The joke is funny (at least to physicists) because most quantum theorists are just about as far from being practical, applied types as it is possible to get. They tend to live in mathematics or philosophy departments rather than physics buildings, regard running a computer simulation as getting their hands dirty, and probably have not been in an honest-to-goodness laboratory since their undergraduate days. You simply cannot imagine them doing the kind of physics experiment that involves spanners and grease.

By these standards, Anton Zeilinger is indeed a quantum mechanic. His excellent physical intuition has enabled him to devise some of the most dramatic experiments to date to test the foundations of quantum mechanics. For example, it was he who first upgraded the two-slit experiment—which caused such excitement when it was first performed with electrons rather than photons—to work with buckyballs, giant molecular cages comprising 60 carbon atoms. Even these football-like structures, each with a definite rigid shape and containing hundreds of fundamental particles, can be demonstrated to be "in two places at once"—or at the very least, exploring two paths at once.

When I first met him, Zeilinger was perfecting a larger-scale version of the Aspect experiment, in which the photon paths could be extended up to several kilometers. He was determined to overcome several potential criticisms of the original Aspect setup. His main goal was to ensure that the choice of polarization measurement for each photon was truly random, because in Aspect's experiments the measurement direction was selected using acoustically driven optical switches that flipped at regular (albeit very fast) intervals, so it was, in principle, predictable in advance.

After describing his new experimental design to a colloquium at Oxford, Zeilinger asked the audience to suggest better ways to make the measurement directions at each end of the experiment completely independent and unpredictable. For example, you could use tables of random numbers generated in advance to set the directions; but it might be better still to decide them only at the very last moment when the photons were in flight, using a real-time random-number generator of a type developed for cryptography. Another alternative suggested was that two human observers, each armed with a toggle switch, could consciously decide the direction of each measurement as suited their whim. Of course I am sure that Zeilinger had already thought of all these ideas, and more, for himself. The purpose of his canvassing the physics community was to make sure that after the experiment was done no one could turn round and say, "Ah, but your measurement choices were not sufficiently random. What you should really have done is *this* . . . ."

Most experimenters who test the foundations of quantum theory are hoping that sooner or later they will turn up something unexpected. After all, for confirming a well-accepted physical theory to an extra decimal place, an experimenter can expect at most a pat on the back. It is discovering something new—for example, something that throws new light on the elusive quantum collapse process—that brings the chance of greater rewards. At a dinner when I had the chance to question Zeilinger in more depth, it became clear that he does not expect the unexpected to turn up at any point—he believes the orthodoxy will be confirmed every time. As we talked, I became increasingly puzzled as to the motive for his admittedly beautiful experiments. Eventually I asked, "Are you trying simply to rub the theorists' noses in the fact that the problems of quantum theory are real, and cannot be ignored as they are demonstrated at ever larger scales?" He beamed and replied, "Yes, that is so. Exactly!"

He has continued to enjoy tweaking the theorists' noses. For example, he designed a new version of the buckyball interference experiment to work with hot buckyballs, ones at such a high temperature that they emit several infrared photons on the way through the two-slit device. He asked various theoreticians whether they expected an interference pattern to be produced. They predicted that it would not, on the basis that the infrared photons striking the walls of the experiment would constitute "measurement"—an irreversible interaction with the environment that would destroy interference. But Zeilinger predicted that he would get an interference pattern after all, because the wavelength of the infrared light emitted by the buckyballs was so long that it did not convey sufficiently accurate information about their position to tell the environment which slit they were going through. And of course he was right.

When I first asked Zeilinger which interpretation of quantum theory he favored, he was reluctant to reply. Eventually he said, "I think there is a need for something completely new. Something that is too different, too unexpected, to be accepted as yet."

"Some variant of many-worlds?" I asked, expecting the answer would be yes. He brought his hand down on the table with a thump and gave a monstrous Teutonic snort.

"No, I do not think many-worlds is right at all. Absolutely not!" And he would not be drawn further. But now, several years later, he has come clean; he does indeed have a radical suggestion, which is new in essence and certainly deserves to be taken seriously. His view is based on one of the most fundamental differences between a universe made up of quantum systems, and one that is classically continuous: A quantum system can contain only a limited amount of information.

Continuous Is Infinite

The difference is rather neatly illustrated by a humorous science fiction story written many years ago by Martin Gardner. In the story, an alien lands on Earth and offers the human race the entire sum of his incredibly advanced race's knowledge, helpfully translated into English. (Apparently Gardner gets a lot of correspondence from people who claim to have had an alien land in their backyard who makes a similar offer. He writes back, politely asking the alien to give the answer to any one of several mathematical problems known to be soluble, although so far unsolved by humans. He never hears back.)

The alien's offer is gratefully accepted. He produces a glass rod from his spaceship. "This rod encodes the entire contents of the Library of Zaarthul," he says. "All you have to do is measure the ratio of its length to its width with an accuracy of 1 billion decimal places. The numbers spell out an English translation in a simple two-digit code where 01 stands for A, 26 for Z, and so on." Then he seals up his saucer and flies off. The human scientists try to measure the length and width of the rod as best they can, but they never get beyond the first few letters of the message.

Gardner was jesting, of course, because glass rods and other physical objects are made up of the quantized units we call atoms. The rod would be about 10^8 atoms wide by 10^9 atoms long, and so its length/width ratio could encode some eight or nine decimal digits of information at most. However, if we lived in a universe where objects were made of a continuous substance that could be subdivided indefinitely, it would be possible in principle to make Gardner's rod. In fact if we lived in a nonquantum universe, it would be positively wasteful for the alien to use such a large physical object. Consider the classical picture of an electron as a tiny spinning top, with its axis of rotation pointing in a specific direction. The alien need hand over only one electron, as shown in Figure 15-1. "Measure the angle between the spin axis of this particle and Galactic North, accurate to one billion decimal places," he says. (Of course "Galactic North" would need to be very precisely defined, perhaps with respect to a master compass consisting of another electron.)

However, in reality, electron spin is a quantum property. The only

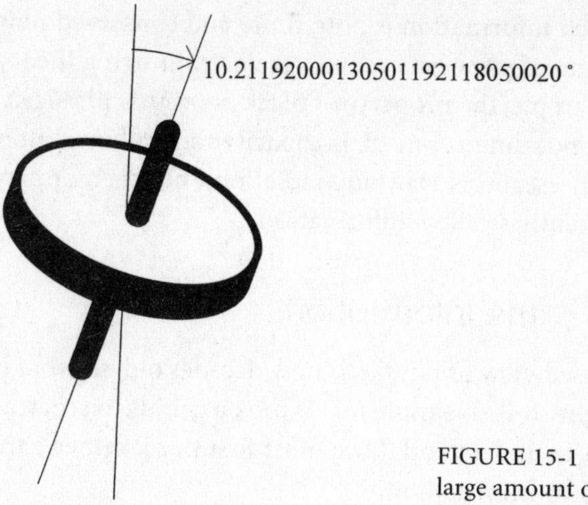

10.2119200013050119211805002 0 °

FIGURE 15-1 An electron encoding a large amount of information.

measurement we can make is whether the spin is up or down relative to an arbitrarily chosen plane. This yes/no answer can yield only a single bit of information and so, in our universe this refinement of the alien's encoding scheme would be not just technologically difficult, but fundamentally impossible.[1] Indeed, modern theory implies that there is an upper bound on the amount of information that any object or system can contain. For a macroscopic object like a glass rod, it becomes very large but certainly not infinite. Correspondingly, only a finite amount of information is required to describe a given object not just approximately but perfectly—to record all the information about it that the universe contains.

This has profound implications for physics. The whole universe we perceive contains only a limited amount of information. It could be described completely by a sufficiently long string of zeros and ones. Our world might therefore be indistinguishable from a digital computer simulation of itself, just as hypothesized in countless science fiction stories. This contrasts completely with a classical, continuous universe for which the simulating computer would have to record an infinite number of digits just to specify the exact position, velocity, and spin of a single fundamental particle.

A picture in which information is both finite and conserved now underpins the thinking of physicists at every scale, from string theory to cosmology. For example, the property of black holes that physicists nowadays find most puzzling is not their capacity to swallow matter (which can eventually escape as Hawking radiation), but their apparent ability to permanently swallow information.

The Information

This information-based view of physics is now decades old, so what is Zeilinger's new insight? It is so simple that it took a genius to see it, as is often the case. But to understand it, we must first think rather hard about what we mean by information.

In information theory, the basic unit of information is the bit; something that can have either of two values. In a computer, one bit is represented by a microscopic switch that can be either on or off. However, a bit, or a sequence of bits, can be interpreted in different ways. For example, a logician tends to think of a bit as denoting the truth value of some proposition, recording whether it is true or false. For a mathematician, the normal use of a bit is as a binary digit. A set of binary digits can represent an integer number, as we saw in the chapter on quantum computing. But the same set of digits could also be denoting a letter in the standard alphabet used to display text characters, or the color of a pixel to be displayed on the screen, or the timbre of a musical note to be played, or many other things. As far as the computer is concerned, a bit is just a bit, a switch that is on or off. But the human programmer can use it in different ways depending on what he is trying to make the computer do. A rival computer programmer, trying to understand what a program is doing by looking at the binary bits it generates, will not get very far until he can interpret what kind of information is being represented by each bit—even though the computer itself could not care less and can function perfectly well without this knowledge.

If we view the universe as a sort of giant computer manipulating a large but finite number of bits, there is still the question of how to interpret the information the universe-computer is storing and processing. The most natural assumption is that each bit of information relates to a particular point in space-time. This is very like the way that computer models of physical systems like the weather work. The difference is that whereas weather simulations on present-day computers have to divide the atmosphere into imaginary cubes measuring kilometers on each side, a true universe-simulation would hold information at a vastly finer scale, of the order of a Planck length.[2]

Zeilinger's approach is radically different. He prefers to think that a given bit of information held by the universe-computer can be interpreted, not as information about what is going on at a specific point of the space-time continuum, but as the logical value (true or false) of statements that can be made about quantum systems. This interpretation allows for the fact that a quantum system considered as a whole can contain information that is not present in its constituent parts.

Nonlocal Information

Figures 15-2 and 15-3 illustrate the principle of distributed information.

FIGURE 15-2 An entirely random pattern.

If you examine either picture on its own, the dot pattern does not merely look random to the unaided eye, it really is arbitrary, and even the best code-breaking machines at the National Security Agency could not extract any meaningful information from it. Yet if you hold the page up to a bright light, you will see a very clear and unambiguous pattern, which of course you are free to interpret as the figure "0," or the Eye of God looking at you, as you please.[3]

How is this possible? We will illustrate with an anecdote. Let us suppose that you are the general in charge of a besieged castle. You want to send a message to your king telling him how many days you can hold out before you will have to surrender if help does not arrive. You have a number of brave volunteers prepared to sneak out at night and try to make it through the enemy lines, which is fortunate because radio has not been invented yet. However, there is a dilemma. You know that if the messenger is intercepted, the result will be disastrous because the enemy will discover exactly how long it has to wait in order to achieve victory.

FIGURE 15-3 Another entirely random pattern.

Then you have a brainwave. One messenger might be intercepted, but if two set out in opposite directions, the chance that they will both be captured and forced to divulge their information is small. You could divide the message crudely, for example, so that one man carries a note saying "One hundred" and the other "and sixteen," but that is not very satisfactory. You really need a way to divide the message so that each note on its own carries no useful information at all, yet both taken together convey the full meaning.

Then the castle mathemagician approaches you bearing a stylus and a piece of parchment. "Sire," he says, "I have a way. The essential problem is that we need to send the king an 8-digit binary number, namely 01110100"

"Quite so," you say, being rather advanced in binary math by the standards of the era.

"Well, I have it," he says. "We will simply send out two messengers, each bearing an 8-digit binary number, and each bearing the magic word "XOR" in the corner. That will tell my colleague Merlin exactly what to do when the messages arrive at the king's castle.

"The number we want to send will be encoded as follows: If a digit in the final message is to be 0, then the two submessages will each have the same digit in that place. If the digit is to be 1, then the two submessages will have different digits in that place.

"Here is how we will generate the submessages. The first digit of the final message is to be 0, so we must write the same digit in both submessages, but we have a free choice whether that digit shall be a 1 or a 0. We will choose by tossing a coin, heads for 1, tails for 0. If I might borrow a coin. . . ."

You give him a gold coin; he tosses it and it lands heads, so he writes a 1 in both submessages.

"The second digit of the final message is to be a 1, so the first digits of each submessage should be different. We will use the coin again. If it falls heads we will insert 1 in the first message and 0 in the second; if tails, 0 in the first message and 1 in the second." He tosses it; it lands tails. He continues in the same way until the strings below have been generated,

Submessage 1	10101001	XOR
Submessage 2	11011101	=
Final Message	01110100	

"Now the marvellous thing is, Sire, that considering the first submessage on its own, each digit was set depending on the toss of a coin. The string is therefore completely random. And considering the second submessage on its own, each digit also depended on the toss of a coin, and it is also completely random. And yet both messages taken together yield the number we want to convey. On receipt of the two pieces of paper, Merlin has only to compare the successive digits, writing down 0 if they are the same in both messages, but 1 if they are different. Why, thank you very much, Sire."

And off he goes, clutching the coin. What he has said is perfectly true. Each number taken on its own is random. Yet their relationship contains a message.[4]

To fully understand the ways a quantum system can contain information, we need to take one further step. The nonlocal correlations we have seen so far each required some information to be held locally, as pixel patterns in the first case and binary numbers in the second. Even though the information in one submessage, or one page of pixels, was not useful to us on its own, it was still information in the strict sense of the word. The remarkable thing about quantum systems is that they are also capable of containing *only* nonlocal information. Fortunately, even this can be illustrated with a classical analogy.

Let us embark on another exotic adventure. This time we will suppose that you are a secret-service agent in a foreign country trying to communicate with a colleague who has been imprisoned locally. Unfortunately the guards will not allow him to be given any kind of object or message, with one exception. Under local custom, some friend of the prisoner is permitted (and indeed required) to pay for his meals by giving the guards two coins, a nickel and a dime, each day. For local cultural reasons, the guards toss the coins in sight of the prisoner, allowing him to see whether they land heads or tails, before they are spent.

This suggests a cunning plan to you. A normal coin does not store any information that can be revealed by tossing it, in the sense that heads are just as likely to come up as tails. However, you discover that by cleverly tampering with the coins you send, you can make them nonrandom; you can make each one always fall heads, or always tails, as you please. If your friend knows this, you can send him a binary message with heads coding for 1 and tails for 0 at a rate of two bits per day, one bit per predictable coin. With patience, a message of any length can be sent.

Unfortunately, disaster strikes. The guards turn out to be not so stupid as they appear. Before taking the coins to the prisoner, they first toss each one a few times out of his sight. If any coin keeps landing the same way up every time, they treat it as suspect and substitute an untampered one. Your scheme is foiled!

Fortunately, you come up with a better one. You start rigging the coins more subtly, perhaps by inserting tiny, cleverly placed magnets. The result is that while each coin individually is equally likely to land heads or tails, the two coins tossed together will always land either the same way up (both heads, or both tails) or opposite ways up (one head, and one tail) depending on how you place the magnets. Each coin on its own contains no information; there is no predicting whether it will land heads or tails on any given toss. However, both coins tossed and viewed together can code one bit of information, say, 0 (if they come up the same way) or 1 (if they come up as opposites). An equivalent coding is to say that your friend should write down a 0 if the logical proposition "The coins have landed the same way up" is true, and 1 if it is false. Now the guards (who are not all that bright) accept your coins as random, and you will be glad to hear that your friend eventually escapes with the aid of the information that you send him at one bit per day.

The idea of a system whose parts appear individually quite random, yet exhibit curious correlations when taken together is no doubt reminding you of something, namely the photons in the Bell-Aspect experiment. Of course the coin correlations are not really spooky, because they occur between objects that are not widely separated, but interacting via well-understood forces. However, it might be instructive to remember that there was a time when the apparent action-at-a-distance effect of a magnet appeared just as spooky to contemporary philosophers as EPR correlations seem to us today.

Zeilinger's Informational Principle

Now at last we are in a position to understand the full flavor of Zeilinger's new hypothesis. Conventionally, the information-carrying properties of quantum systems are derived from fiercely complicated equations. Zeilinger's approach is to assume *as an axiom* that the amount of information the universe holds about a quantum system is finite and bounded. In his view, an experimenter who tries to measure incompatible information about a quantum system is making the same kind of mistake as a rookie computer programmer who tries to read 16 digits from an 8-digit register. The extra information simply is not there—anywhere. His insight can be applied straightaway to the most basic demonstration of quantum properties, the two-slit experiment. We know that if we fire a succession of photons or other particles through two adjacent slits, interference will normally produce a pattern. The pattern develops slowly; a clear interference-band picture requires many bits to define it. If you watch the pattern build, it is much like downloading a picture from the Internet through a slow modem. The first few hundred bits give only a blurry view, which becomes gradually sharper as more bits are transmitted, as shown in Figure 15-4.

But if any attempt is made to measure which of the two slits each particle passes through, however delicate or indirect the means employed, the interference pattern is destroyed, as in Figure 15-5. In Zeilinger's view, this is because each particle carries just one bit of trajectory information. We can use this bit either to get which-slit information or to increase the definition of the picture on the photographic film, but not both. If we measure the trajectory of every particle, because it takes exactly one bit (coding, say, 0 for left and 1 for right) to specify which slit, there is no capacity left over to code picture information, and your film will show a random pattern of dots. If you measure only, say, half of the photons, the pattern that builds up

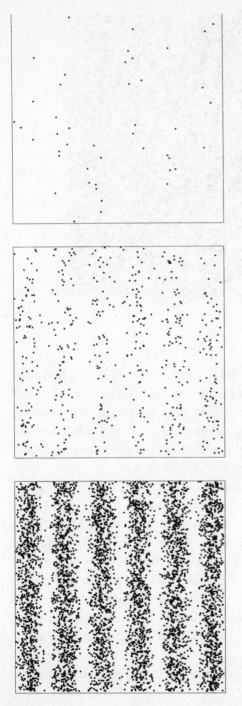

FIGURE 15-4 Two-slit interference pattern.

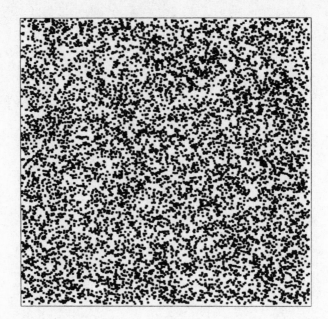

FIGURE 15-5 No pattern.

will be blurred as in Figure 15-6, because only the nonmeasured photons can contribute picture information. Each bit can be used only
once; trying to obtain both trajectory information and interference-
picture information from a limited number of bits is much like trying
to use the same area of computer memory for both numerical and
picture data—something inevitably gets corrupted.

Zeilinger's view seems to imply that much (or perhaps even all) of
the information the universe-computer contains is relational in nature—it can know the relative status of two variables, without storing
any information about their absolute values. In terms of our parable
of the besieged castle, the computer knows the contents of the overall
message to Merlin—but it holds no data about the individual
submessages carried by the two runners. This leads Zeilinger to the
idea that the fundamental information in the universe-computer
should be regarded as logical true-false values of statements about
quantum systems.

Zeilinger has proved that properties of quantum systems that are

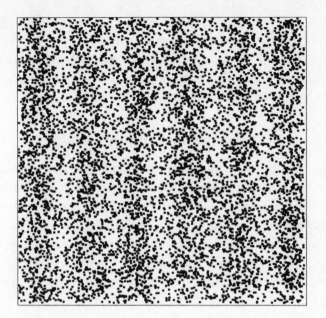

FIGURE 15-6 Partial pattern.

often considered weird, like the correlations obtained in nonlocal measurements, follow logically from this principle. His starting point is a simple system, the two photons of the Bell-Aspect experiment. Zeilinger finds that the universe-computer holds only two bits of information to describe their joint polarization, measured at whatever angles. These two bits can be considered as the truth values of the two statements,

"The polarization of the two photons, measured in parallel directions, will be the same." (Always TRUE.)

"The polarization of the two photons, measured at right angles, will be the same." (Always FALSE.)

For this system, all that the universe-computer contains is relative information. There is no information capacity left to store the states of the individual photons. Zeilinger finds that from these assumptions, he can recreate the spooky correlations of the Bell inequality. He goes on to derive a more general result, which, exceptionally for this arcane field, can be rewritten in simple English: "Spooky correlations can arise in a simple quantum system when more than half of the available information is used to define joint properties."

Thus the two-photon system of the Aspect experiments turns out to be, surprisingly, a lot more quantumy than the minimum necessary for Bell correlation effects to occur.

An Informational Interpretation?

Zeilinger has certainly found an interesting new way to look at entanglement. His success in explaining nonlocal behavior from straightforward assumptions is solid Occam's-razor justification for his hypothesis that, at the most basic level, the universe might contain information about individual quantum systems rather than individual localities. He presumably hopes that his approach can be extended mathematically to determine the behavior of more complex entangled systems. If this were to throw light on the way that relative information in small systems tends to "turn absolute" in larger ones, it could provide a new way to look at quantum collapse. Unfortunately, previous attempts to extend such "measures of quantumness" to large systems have run into a morass.[5]

Zeilinger's view also shares a problem with much less worthy attempts to brush aside the problems of quantum, namely the questions: If the universe is intrinsically nonlocal, why is the *illusion* of locality so strong? Why do causative effects always propagate at less than the speed of light? Why are forcelike interactions stronger at short ranges? Nevertheless, if it turns out to be possible to generate further real physics from an extension of his axioms, his ideas will have to be taken very seriously. Perhaps it will turn out that quantum is the real stuff, and the illusion of locality arises as an almost incidental feature of the algorithm the universe-computer is running.

Personally, however, I do not believe that Zeilinger's approach will lead to the best way to understand the quantum world. When we discussed the merits of different interpretations in the context of tick-tack-toe, we decided that it was vital to find a game that humans are intuitively able to play. In terms of the tick-tack-toe analogy, Zeilinger's method is like trying to play the adds-to-15 game. Our human minds are designed to perceive the world in a visuospatial way more easily than in terms of abstract logic. However mathematically successful Zeilinger's approach turns out to be (and it still has major obstacles to overcome), we would still need the equivalent of a magic square to translate his informational universe into one we can readily visualize.

CHAPTER 16

PROVING AND
IMPROVING MANY-WORLDS

At the very least, we have established that the many-worlds view is a valuable thinking tool, worthy of its place among interpretations of quantum. Certainly it is the best way to think about the interaction-free measurements described in Chapter 10. With the modest principle, "Interference effects between worlds persist until a measurement of the self-interfering object is made in either world," we were able to understand the workings not only of the Elitzur-Vaidman and Zeilinger designs, which use photons following different trajectories from the one in our "own" world, but more subtly the Paul-Pavicic monolith experiment that uses a photon that left at a different time than the one in our own world.

I suspect that even considered just as a conceptual model, many-worlds has a great deal of further mileage waiting to be wrung out of it. For example, no one has yet given a clear, simple picture of why particles can "quantum tunnel" forward faster than light, yet not carry information faster than light. The idea that we might (loosely and poetically speaking) have swapped the particle that left a moment later in "our" world for one that left just an instant earlier in a not-yet-decohered other world is at least an interesting try; the principle above illustrates why tinkering with either particle would have broken the connection, ensuring no message could hop ahead faster than light.

The idea that when we make a measurement on an entangled system we are in some sense "dialing in" to a world in which other parts of the system are likely to have certain values relative to our own explains how we can later find we have obtained spooky correlations with far-flung parts of the system, without any question of being able to cause effects on those parts. (The dialing-in metaphor must be qualified by a restriction like the rule that arrested persons can make one phone call only before being isolated in a cell. You lose even this weak spooky link once you have used it.)

It seems quite likely that extending this "dialing in" or "mix-'n-match" rule of thumb might help visualize other features of entanglement, for example, giving us better insights into such phenomena as so-called quantum teleportation. If a richer view of many-worlds gives us better insights into how to design such things as quantum computers, the imaginative effort will have been well worth it.

Other worlds are at any rate a useful illusion. But is there any hope of demonstrating that many-worlds is more than just an interpretation, or to put it another way, that those interpretations that do not accord other worlds equal status to our own are falsifiable?

David Deutsch and Lev Vaidman have each proposed hypothetical experiments that would "prove" many-worlds by demonstrating interference with world lines that we would normally think of as having irreversibly decohered from our own. For example, if we could replace the photon or electron normally fired through a two-slit experiment with a capsule containing a conscious observer, the observer might show signs of having been "affected by" both worlds when she emerged from the experiment. She might, for example, say that she clearly remembers that she could see which of the two paths she was going along—but somehow cannot now remember which it actually was.

These experiments would require incredibly advanced and elaborate technology, however. This has two disadvantages. The first is that no such technology is available, nor will it be for the foreseeable future. The second is what I call the "Jurassic Park" argument. Imagine that two scientists are arguing furiously about whether dinosaurs still exist; they make a large bet on the matter. One guy goes off and by mining Antarctica for deep-frozen dinosaur DNA, etc., he eventually produces a dinosaur. Did he prove that dinosaurs exist or just that it was possible to reconstruct one by heroic feats of data retrieval? In the same fashion a skeptical single-worlder might say that Deutsch's and Vaidman's experiments prove nothing more than that you can create a kind of artificial dual–world simulation by creating circumstances that would never arise naturally.

David Deutsch has claimed that the feasibility of quantum computers pretty much "proves" the reality of many-worlds because where else can the resources for all that computation be coming from? So far, unfortunately, any proof that a quantum computer can fundamentally outperform a classical one remains elusive. And even if that proof is obtained, it will, arguably, show only that our world possesses certain extra degrees of freedom—that what Gell-Mann calls weak decoherence can occur—rather than the existence of strongly decohered world lines effectively independent of our own. There is always an understandable temptation for proponents of any particular quantum interpretation to see stronger evidence for it than a particular experiment provides. For example, shortly before this book went to press, an ingenious experiment by Shahriar Afshar was claimed to have "disproved" both the Copenhagen and the many-worlds interpretations.[1] In fact, it is exactly consistent with the modern many-worlds view, specifically the idea that interference effects from "other-worldly" photons continue up to the point where a measurement is made. Afshar's experiment demonstrates wavelike behavior followed by particle-like detection, just like our bomb detectors.

⬤⬤⬤

But how satisfying it would be if we could directly prove the existence of other worlds! Max Tegmark has one idea for a relatively low-tech experiment to do so. It is simply an iterated form of the quantum Russian roulette idea we have already met. The idea is that you rig up a kind of machine gun that fires one shot per second. However, each second a quantum randomizing device, the equivalent of a coin toss, determines whether the gun will fire a live shot or a blank. You could use a photon that is reflected or transmitted by a partly silvered mirror. If the photon is transmitted, the gun fires a blank bullet; if reflected, a live one.

Having set up the device, you give it a test run. You can be confident that the usual laws of statistics will be obeyed—after 100 shots, there will be approximately 50 bullet holes in the dummy target you have set up. You will hear a sequence something like: Click-BANG!, click, click, click-BANG!, click-BANG! . . . .

But now you step in front of the device yourself: Click, click, click, click, click, click, click. . . . Now you observe a blank every time. Each time the gun operates, you are halving your measure of existence. But of course, you are not aware of those worlds in which you have just ceased to exist. To you, it seems you are invincible. After 10 clicks, you know your chance of survival in a classical world is just less than 1 in a 1,000. After 20 clicks, 1 in a million. After 30, 1 in a billion. At any time, you can prove to yourself the device is working just by stepping aside. Immediately the laws of chance (as seen from your point of view) return to normal. The intermittent live shots start up again.

Tegmark points out a snag with the device (that is, a snag over and above the reservations about quantum suicide listed in previous chapters). He argues that you can convince only one person by the method, namely yourself. Suppose your colleague Professor Cope heroically volunteers to take the stand. You can be virtually certain that after a few shots you will be looking down in horror at the body of the famous physicist.

However, here Tegmark has perhaps not considered quite the whole story. To see why, suppose you invite your secretary in to witness the procedure. "Don't worry," you tell her mendaciously, "I have figured out a clever reason why you will see me survive every time." And you start the device.

Now in most of the resultant worlds, your secretary will very shortly be shaking her head in sadness but not in surprise as her opinion of her boss's sanity is finally confirmed. But those worlds do not matter to you. In the worlds that do matter, she (and any other witnesses you might have invited, such as philosophers of science) are gazing at you with an increasingly wild surmise. Those who know their physics know that the laws of quantum do not explain the miraculous sequence of luck they are seeing. And yet it is happening before their eyes. If you are unscrupulous enough to claim that your survival is due to divine intervention, soon quite a lot of the people in the world you end up in will believe you. As you continue to survive every time, despite the most rigorous checking of your equipment by independent experts, even the most hardened skeptics will begin to wonder. . . .

In fact an analogous procedure was suggested in a detective story written decades ago. The basic idea was to send letters to a large batch of randomly selected people claiming that you have inside information and can predict the result of Sunday's big game. But actually half of the letters you send predict side A will win, the other half side B. After Sunday, you discard whichever half of your address list you sent wrong tips to, but write to the other half a second time with a new prediction for next week's match. Again you split your prediction, and are left with a quarter of the original batch that is beginning to believe you. After a month, the small batch remaining is convinced that you know what you are talking about, and most of them have profited from your knowledge by placing bets. Now you tell them that your infallible tipster service will continue, but the annual subscription is $10,000 payable in advance. . . .

In the original story, the protagonist unintentionally creates a dangerously fanatical cult of people who believe in him more strongly than he ever intended. That story was written before the days of the Internet, but in the present e-mail era the scam would be perfectly feasible. (Come to think of it, some of the financial-advice spam I get could quite possibly be generated on this principle.) In the quantum-suicide version, however, you end up not with a handful of people to whom you appear infallible, but a whole world.

❧❧❧

It would obviously be desirable to have a less dubious method of proving many-worlds. In the mid-1990s, Rainer Plaga of the Max Planck Institute proposed a less dangerous experiment.[2] The idea is to first create a miniature Schrödinger's cat, an ion in a state of quantum superposition, and then do a separate measurement that causes a clearly signaled world-split, for example, by firing a photon into an apparatus that lights either a green lamp or a red one, depending on whether a photon is reflected by a half-silvered mirror. Plaga's reasoning is essentially that because the ion in its magnetic cage does not yet know whether the green or the red lamp lit, it remains in effective contact with both worlds. Thus a measurement interaction triggered in one world—by firing a laser beam at the ion, for example—could cause an effect in the other as the quantum superposition is destroyed. For example, it could cause an electron to be emitted at that moment.

If the experiment works, it can be used to convey information from one world to the other in a one-shot kind of way. Here is how it could be used to convince a many-worlds skeptic. Ask him to invent a six-digit number unknown to anybody else and lock it in a safe to which he holds the only key. Explain that the rules of the experiment are that just before midday, a button will be pressed on the green-or-red-lamp device. If the green light shines, he must open the safe and tell you the number. But if the red light shines, he can keep the safe locked—and just after midday, you will tell him the number, transmitted as a signal from the other world where the green lamp shone. The experiment is run; the red light shines. To the skeptic's astonishment, a few moments later you can tell him his secret number, even though it is still locked in the safe.

Here is how the trick is done. If the green light shines and the safe is opened, you read the number and set a timer that causes the Schrödinger's-cat ion to be interrogated by a laser beam at exactly the number of microseconds after midday corresponding to the value of the six-digit integer. In the parallel world where the red light shone, the ion emits an electron as its twin is interrogated. By measuring the exact time at which the emission takes places in microseconds past midday, the code number is discovered. Of course half the time you do the experiment, you will end up in the world where the green light shines, and become the transmitter rather than the receiver of the information. But all you have to do is keep repeating the experiment. In half of the runs, on average, you get to show the skeptic information that is known only to himself and persons in the other world.

This would be a repeatable and utterly convincing demonstration of many-worlds. Unfortunately, very few physicists think the experiment would work—Plaga himself put it forward only tentatively. The overwhelmingly majority view is that the worlds would completely decohere at the moment the green and red lamps lit, including divergent versions of the Schrödinger's-cat ion. To the best of my knowledge, the experiment has not even been tried in the decade since it was proposed.

∽∾∽∾

Last night I had a dream. . . .

I was sharing a lab office with Professor Cope, the impressive old gentleman we met in Chapter 1. This was proving to be quite a trial. As you may remember, Cope had at first been most reluctant to accept the many-worlds theory, but now he was insisting on telling me about his new scheme for transmitting messages between worlds that had completely decohered and gone their separate ways. I managed to tune him out as he told me the details (I now regret to say), but ignoring the banging as he constructed an odd-looking device to be connected to his computer was more difficult. Presently he turned to me with an expectant air.

"Congratulate me!" he announced. "I have connected the camera on top of my workstation so that it will transmit a picture to another copy of my computer screen in a parallel world, and vice versa. Words I speak into my microphone will also emerge in the headphones of my other self in that adjacent world!"

He pointed at his workstation, which was displaying a picture of himself. He waved, and the image mirrored the action.

"But now watch!" he said, throwing a bulky switch. "That simple action established communication between two diverging worlds, in one of which a photon was reflected, in the other transmitted." He waved his arms about. The picture on the screen continued to copy him exactly. He looked mildly disconcerted.

"Harry," he said loudly, presumably trying to speak to the other version of himself, "*you* raise your hands above your head, *I* will hold mine out to the sides." He held his arms out to the sides and the image on the screen duplicated his action exactly. It was obvious that his experiment was a fiasco. He was merely continuing to see his own picture in one world. Eventually he went home in a bad temper, leaving the computer turned on.

The following day I arrived at the lab to be greeted by a sarcastic voice from Professor Cope's computer. "Well, I see that *you're* on time," it said.

I looked at the screen. It seemed to be showing Professor Cope sitting in his lab chair, but the real chair was empty. The image grinned. "That's right," it said. "I'm the Cope in the world next door to you."

"Very funny," I replied. "I suppose you're sitting at your computer at home, getting some fancy graphics software to display you with a background of the lab here. Well, it's not April First, and I'm not fooled."

The image shook its head. "No fooling," it said. "I'm a little ahead of your Professor Cope. In my world an insect blew into the window early this morning, and the bang woke me. Your version seems to have slept in."

At that moment the lab door opened and Professor Cope himself walked in. "Beat you!" called the image in the screen. Cope looked at it and appeared genuinely flabbergasted.

"I got into work a couple of hours hour ago, so I've had more time to think about all this than you," the image said to the physical Cope cheerfully. "Of course when our worlds started to diverge yesterday, they were incredibly similar, with only one photon's worth of difference between them. No matter what the two of us did, we couldn't help acting exactly like the same person. But as time passed, butterfly effects magnified the difference so that we started to act differently. Now there really are two of us, with completely different thought patterns.

"Pull up a chair and listen to the plans I've figured out. This thing is incredible—we're going to be *rich*."

And then, of course, I woke up. Communication between parallel worlds is fun to explore as a science-fiction theme. In my opinion the consequences that would follow have not yet been worked out as thoroughly in contemporary science fiction as other hypothetical scenarios—time travel, matter transmitters, and so forth—were explored back in the golden age of sci-fi. Perhaps one day I will yield to the temptation. But on almost all present indications, parallel worlds that you can interact with remain the stuff of pure fantasy.

It is too soon to be absolutely dogmatic about this. We do not yet understand everything about decoherence, or about the relationship between quantum and general relativity, to name just two areas. It has even been suggested that many-worlds could solve a troubling paradox of modern physics—that general relativity implies that at least one method of time travel is possible. In a single world line, obvious contradictions could arise if you try to change your own past. But from a many-worlds perspective, all you would be doing is creating or entering (depending on which way you look at it) new measures of world lines, diverging from those that produced your original memories.

∞∞∞∞

A more promising line of approach, however, is to strengthen the philosophical case for many-worlds. Let us start by further disparaging the idea that the supposed extravagance of the many-worlds view is a reason for rejecting it.

Deutsch in particular has pointed out that the many-worlds interpretation is very like Bohmian mechanics—the particle-plus-guide-wave idea we followed at the start of the book—*minus* the idea that there are particles riding the waves in just a few particular positions. As he points out, if we accept the reality of the waves, why ever should we assume that all but a few positions on the "sea" are empty? Lev Vaidman has put it more poetically:[3]

"If a component of the quantum state of the universe, which is a wave function in a shape of a man, continues to move (to live?!) exactly as a man does, in what sense it is not a man? How do I know that I am not this 'empty' wave?"

Of course, Bohmian mechanics is not the only alternative to many-worlds. But again, as Deutsch has pointed out, other approaches that allow for some kind of local-fixed-reality are actually even more extravagant. If we forget the worries about backward-in-time paradoxes, and assume that when, for example, we test one photon in a Bell-Aspect experiment it really does communicate with the other via a faster-than-light signal, just imagine how many such signals must go to and fro. Every time a particle decides what to do, it must consult with all the other particles that it has ever interacted with (and therefore to some degree become entangled with), which in turn must consult with all the particles that *they* have ever interacted with, and so on. We must postulate an absurd amount of behind-the-scenes messaging going on, which at least rivals the supposed extravagance of the multiverse.

But now let us take a more aggressive approach. Let us demonstrate two possible ways to wield Occam's razor very strongly in support of the many-worlds view. The first is based on an insight of Max Tegmark's; the second is my own.

Although we only have one universe to examine, certain of its features are very striking. In particular, the physical laws defining its behavior are remarkably few and algorithmically simple—they can be written on a single sheet of paper. Its starting condition, essentially as a single point, was also simple. This conforms to our intuitive expectation—although perhaps it generates our intuitive expectation—that a universe that can be defined by a small amount of information, however large the volume of space and time it might eventually expand into, is much more likely than a universe embodying a vast set of rules or a quirky set of initial conditions that would require a great deal of information to describe it .

Max Tegmark has identified a troubling problem with this cosy "universe from a tiny package of information" view.[4] The universe that we see around us contains a mind-boggling amount of detail. The general pattern of the universe that we see can be explained, we now understand, by the phenomenon called self-organized complexity. Every region of the universe—and indeed of any universe whose rules are sufficiently similar—will contain stars, galaxies, complex organic chemicals that give evolution a potential starting point, and so on. But we also see a great deal of specific detail that cannot be explained by such general rules. Why is the play Hamlet worded exactly as it is, for example, and written in an alphabet using 26 characters? Although the text of that play, like almost any long text written in the English language, can be compressed by a factor of several using clever com-

puter algorithms, the irreducible amount of information it contains—
in effect, the length of the shortest computer program that could reproduce the play exactly—is still of the order of 100,000 binary digits.
To describe the state of even the planet Earth and its contents exactly,
let alone that of the whole visible universe, would take an enormous,
perhaps infinite, amount of information. Where did all that information come from?

Tegmark has a simple answer. If we live in a multiverse in which
every physically possible quantum outcome occurs, the detail is merely
a kind of observer illusion. There are equally valid universes in which
the play Hamlet, for example, takes many slightly different forms. And
each contains observers who wonder why it took exactly that form. It
takes less information to specify a multitude of possibilities than it
does to specify a single possibility. To write down a specific sequence
of the result of tossing a classical or quantum coin a million times
requires a million binary digits. But to tell you that the result is $2^{1,000,000}$
equal measures of universes in which each of these sequences occurs
takes just a single sentence.

Although Tegmark does not use the metaphor, there is a hypothetical library that philosophers are fond of invoking, which sums up
his idea very well. It is a library containing every possible book that
could ever be written, and yet no useful information! Figure 16-1
shows a simple computer program, storable in fewer than 1,000 binary digits, that can generate the exact text of Hamlet. In fact it can
generate alternative versions of Hamlet that are better than
Shakespeare's, and indeed the text of every other book that was ever
written or can ever be written that is composed only of standard English letters and the common punctuation marks and less than about
a million words long.

The program shown really could—at least, given a very durable
computer and a very large supply of paper—generate the hypothetical
library the philosophers are fond of describing. The first iteration of
the program will simply print out in sequence every possible book
that is only one character long—about 100 of them if we allow upper-
case and lowercase letters and punctuation marks. The second itera-

```
    START
    FOR  BookLength = 1 TO 7000000
    CALL  GenText(BookLength, "")
    NEXT  BookLength
STOP

SUB  GenText(Lremaining, S$ )
  IF  Lremaining = 0  THEN
    PRINT  S$ + "~~~End Of Book~~~"
    ELSE
      FOR  I = 20 TO 120
        CALL  GenText(Lremaining - 1,
S$ + CHR$(I))
      NEXT  I
    END IF
END SUB
```

FIGURE 16-1 A program cleverer than Shakespeare?

tion will print out around 10,000 books containing every possible pair of characters, and so on. The library produced will be exhaustive, but it will not be very useful. But it does make Tegmark's point rather well. If a particular Hamlet could arise from the text of a particular version of the play in the library and ask "Why should the sequence of events just described happen to *me*?" the unsympathetic answer is, "That's just the way it looks to you. Actually, every typographically describable sequence of events happens to some equally real Hamlet somewhere in the library!"

The multiverse generates every physically possible sequence of events simultaneously—and requires very little information to set it going, just as the book-writing program above is very short. Surely it is more plausible that we are just one of many sets of creatures living in a universe that requires little information to describe it, than that we are a unique set of creatures living in a universe that requires a lot of information to describe it?

Constructing a Local Universe

One great advantage of a multiverse as a visualization tool should be its locality, the avoidance of the need to postulate instant long-range influences. In an earlier chapter we mentioned David Deutsch's key paper which proves this, and we described some metaphors like "dialing in" to a particular world when you measure an entangled system. But so far we have not really gotten the full benefit of multiverse locality in a way we can feel in our bones. As the philosopher Jim Cushing put it, we need to tell ourselves local stories in order to feel that the universe is working in a commonsense way. We need a story in which space is filled with entities that have effects only on their immediate neighbors, and in a well-defined temporal sequence. The universe I am asking you to visualize is, of course, a hidden-local-variable theory. And I would suggest, very controversially, that Deutsch's result tells us that such a thing is possible—that it is a legitimate board on which to set out to play tick-tack-toe with the gods, a potentially valid way to look at things. A hidden-local-variable multiverse theory can work where a hidden-local-variable single world line cannot. We know that this is possible in principle. Have we any way to put some flesh on the bones, to deduce the properties of the postulated unobservable cogwheels whose turning supports the persistent patterns we can see?

The first clue, of course, comes from the fact that the hidden variables are indeed hidden, not directly observable by us. This feature is much more disconcerting to the layperson than to the physicist, because physicists know that if you are anywhere—be it a universe or a multiverse—in which the laws of physics operate in a time-symmetric way, with things bouncing about elastically, there is no reason to expect that any observer should be able to see and record all the goings-on of the variables. On the contrary, making any kind of persistent pattern or "permanent" record is a rather rare and special process.

We can actually find classical systems in which a single region of space supports many independent processes that hardly interact. An earlier illustration we used featured a man-made object of this kind, a computer made of optical fibers through which different wavelengths of light are transmitted by deliberately contrived arrangements of filters. More convincing would be an example of a volume of empty space that is shared by largely separate processes.

Allow me to introduce you to the Radioheads. They are creatures living in interstellar space who are so tiny, or so ghostlike, that they never affect one another by direct physical contact. They perceive one another because each has a little built-in radio transmitter set to a very precise frequency, and a receiver set to the same frequency. Thus the initial population of Radioheads can all perceive and talk to one another.

Alas, mutation does its work. Each baby Radiohead is born with a receiver-transmitter set to a very slightly different frequency from that of its parents. Although all the parents can see and talk to their offspring, and vice versa, some of the offspring can perceive each other only dimly. As the generations pass, it comes about quite naturally that any one Radiohead perceives only a tiny subset of the total population. As far as he is concerned, most of his distant cousins have passed into invisible ghosthood, their only impact on his existence a faint hiss of background noise. The descendants have split into different species that will never again reunite. They have decohered.

Admittedly I invented the Radioheads. But there is at least one natural cosmological process in which different entities can share the same volume of space with relatively little interaction between them. That is the situation where two galaxies or clusters of stars collide at high speed. Actually "collide" is a complete misnomer for what takes place, because stars are such tiny things in proportion to the vast gulf of space that typically separates stellar neighbors that the chance of physical collision between pairs of stars is virtually negligible. The galaxies pass through one another and continue on their separate ways.

The only interaction is gravitational. You might expect that to be rather significant. In our galaxy, our Sun is more than four light-years from its nearest reasonably massive neighbor, Alpha Proximi. If another similar galaxy were to pass through ours at high relative speed, a number of its stars would be statistically likely to pass the Sun at just a fraction of that distance. However, the direction and magnitude of the tiny gravitational force that Alpha Proximi exerts remains roughly constant over thousands of years, adding up to a significant effect. By comparison, the gravity from those stars of the other galaxy that passed near the Sun would exert forces only briefly, and in essentially random directions. Galaxies that pass through one another at high relative speeds have relatively little gravitational effect on one another.[5]

Allow me to promote you to godhood. You are in charge of creating a universe with three dimensions of space and one of time, just like our own. You can populate the universe with whatever kinds of fields and other things you like, as long as they interact only locally. For bonus points, the rules of your universe should permit interesting patterns to arise.

I put it to you that in general there is no reason to expect that your rules should cause every pattern to continue to interact with every other pattern. That is actually a rather special case. In biology, the laws of evolution naturally cause species to split into subspecies that can no longer interbreed, and diverge thereafter. In chemistry, there are reactions so specific that two or more different chemical processes can be taking place in the same test tube while having virtually no effect on one another. In physics, two water waves can pass through one another and each continue on its way almost unchanged. In just such a way as these, patterns that interact only with a small subset of all the other patterns around should be considered the norm rather than the exception.

Lattice Models

What specific facts can we deduce about our hypothetical hidden variables? Here we must make a diversion into a different area of physics. It is a still-emerging field, the arena of strings and loops and related theories. Not just the fine details, but even their most basic paradigms—the number of dimensions and the very topologies—of the entities involved are still being hotly disputed. But the basic aim was summed up by Richard Feynman when he wrote:

> It always bothers me that, according to the laws as we understand them today, it takes a computing machine an infinite number of operations to figure out what goes on in no matter how tiny a region of space, and no matter how tiny a region of time. How can all that be going on in that tiny space? . . . . I have often made the hypothesis that ultimately physics

will not require a mathematical statement, that in the end the machinery will be revealed, and the laws will turn out to be simple, like the checkerboard with all its apparent complexities. [6]

Caricatured in the simplest terms, string theorists are looking for a model of the universe that will be like a cellular automaton, with space divided into cells, each of which contains one bit of information, and which evolves according to simple local rules a little like John Conway's famous game of Life, as shown in Figure 16-2. The rules of Life are very simple: place counters on a checkerboard. Each turn, remove every counter that has fewer than two or more than three neighbors, but place a new counter on any square that has exactly two neighbors. These rules turn out to be capable of supporting processes of unlimited complexity. Figure 16-2 shows a simple Life position progressing through successive time steps.

Almost no one expects that the fundamental structure of our universe will turn out to be something quite so simple as a cubic lattice with each cube containing one bit of information, as a naive extrapolation from Life would imply. The simplest plausible topology is something like that shown in Figure 16-3, where space is described by some kind of continuously morphing network of locally connected vertices; to conform with special relativity's prediction that there is no special frame that can be considered stationary, the vertices would not be static, but would vibrate about at the speed of light.

There are many other possibilities. But an essential feature of all these models is that rather than every region of space containing an infinite amount of information—as would be required, for example, to define the exact value of a classical field at every point throughout the region—a given volume of space, say 1 cubic centimeter, requires only a finite number of binary digits to describe its state precisely. Put another way, it would be fundamentally impossible to store more than a certain number of bits of information in a region of space of given size.

But how many bits? What is the notional volume required to store a single bit? Presumably it must be very small. An early guess was based on a unit called the Planck length. Human measures of length like the foot and the meter are arbitrary (the true origins of the English foot

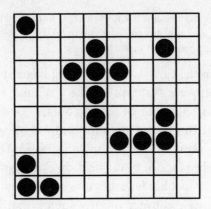

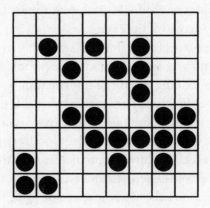

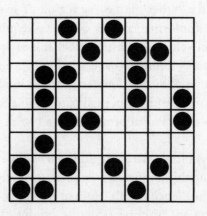

FIGURE 16-2 John Conway's Game of Life.

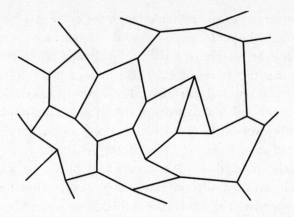

FIGURE 16-3 The idea that the fundamental particles of physics are merely topological features or knots in the fabric of space-time dates back at least to the Victorian notion of ether vortices. But the precise nature of the entities involved continues to be argued. Are we talking one-dimensional strings or two-dimensional membranes, and embedded in a space of how many dimensions? This picture is almost certainly an oversimplification.

are lost in the mists of time; the French meter represents a slightly inaccurate guess at 1 forty-millionth of Earth's equatorial circumference, intended to make navigational calculations easier). More fundamental units are those based on the constants of nature, the most familiar of which is the speed of light, usually written as c. If you told an alien in a radio message that in Britain, autos are restricted to a top speed of 70 miles per hour, he would have no idea how fast that was, but if you told him the speed limit was one 1-millionth of the speed of light, that is a universally meaningful measure.

Another such fundamental value in our universe is the gravitational constant, defining the warping of space that a given mass will induce. And a third is Planck's constant, which we met earlier and which defines the ratio between the frequency of a photon of light and the amount of energy it carries. By appropriate multiplication and division we can derive the basic units of mass, length, and time from these values. The fundamental unit of length, the Planck length, turns out to be tiny, roughly 10^{-35} meter (for comparison, a proton is about 10^{-15} meter in diameter).

There is a very hand-waving argument that the fundamental information density of space should be on the order of 1 bit per cubic Planck unit. In 1973, this guess received a curious kind of confirmation. Jacob Bekenstein discovered,[7] in work later refined by Stephen Hawking, that the region of space containing a black hole, an event horizon, has a physical quantity called entropy associated with it, which in turn implies a quantity of information. By simple thought experiments involving general relativity (for example, considering the viewpoint of an observer who is in normal space, but accelerating), it can be demonstrated that not just a black hole, but any region of space, can contain only a finite amount of information. But there was a shocking surprise. The amount of information any region of space, however shaped, can contain is proportional not to its volume but to its surface area!

This result has been dubbed the holographic principle. The Nobel Prize winner van 't Hooft has given a memorable way to visualize this. If you imagine the surface enclosing a region of space as a flexible computer screen, each of whose pixels is exactly 2×2 Planck units and can be either black or white, then the surface of the screen encodes all the information that region of space contains.

Of course this is all very counterintuitive. If region A contains amount of information X, and region B contains amount of information Y, then surely joining the regions should give us a storage capacity $X+Y$? But Bekenstein's bound tells us that the sum is always less than this. For example, a cube 1 centimeter on a side, the size of a sugar lump, can store approximately 10^{66} bits; but a crate 1 meter on a side containing a million of those cubes can store only 10^{70}, rather than 10^{72}, bits. Where did all the extra capacity go? I have heard van 't Hooft, among others, admit that he finds it very baffling.

But suppose that the universe does consist of information at a density of about 1 bit per cubic Planck unit at the finest scale, as Bekenstein's rule would seem to imply, and that this information evolves via local interactions. What would we expect to observe? Everything we know about the laws of physics gives us a strong hint that the rules of the Planck-scale interactions will be reversible, at least to a good approximation: There will be no intrinsic arrow of time. This means that we cannot possibly expect to store or retrieve information at this scale: The bits will be flickering from one value to another much too unpredictably. They would represent a kind of subinformation that we would not expect to be able to access directly.

We would expect stable patterns—accessible bits of information as used by IGUSes like ourselves, which can be written, remembered, and read—to exist at best as correlations between the Planck-level bits. Figures 15-2 and 15-3 give us a crude visualization: The pixels printed on each side of the paper represent subinformation, but the pattern which is revealed by the process of comparing them (in this case when the page is held up to the light) contains "real" information. Note that this real information is being stored nonlocally: You could slice the page in two with a sharp razor and take the two sides far apart; then the real information could not be said to be contained in either piece on its own, but it is still present.

If there is anything to my speculation, in reality it probably takes not just two but many bits of subinformation to store one bit of "real" information. A better metaphor will be familiar to all readers, although it usually goes unnoticed: the column of light switches found on a typical stairwell. Using a simple trick invented by the Victorians (no modern electronics is required) the switches are wired up in such a way that toggling the switch on any floor switches the light at the top between on and off, irrespective of the current positions of the switches on all the other floors. Here the position of the switch on each floor, up or down, represents 1 bit of subinformation; the state of the light, on or off, represents 1 bit of real information, a property of the whole system.

Now let us return to the multiverse picture. If a multiverse-computer has a storage capacity of 1 bit per cubic Planck unit, and supports a multiplicity of the reasonably stable entities we have dubbed local worlds, obviously not all of the worlds can make independent use of the same storage. A problem will arise rather like that of sharing a finite amount of radio waveband between different users; the more transmitters, the more unavoidable cross-talk there is as each extra user contributes to the general background of noise. The amount of multiverse-information a region of space can store does indeed increase with the cube of its linear dimension, but if the number of stable processes in which that region participates also increases—say, in proportion to its linear dimension, the time light takes to cross the region—then that explains why its available information storage capacity, from the point of view of any one world process, increases only with the square of the dimension, just as we observe. Perhaps a cube 10^{100} Planck units on a side can indeed store about 10^{300} bits of subinformation or multiverse information, but this capacity has to be divided between 10^{100} world processes, giving by simple division only 10^{200} bits of stable "real" information capacity available to each.

I emphasize that this picture of a hidden-variable multiverse is very speculative—I am taking the license traditionally allowed an author in the last chapter of a science book to its limits! But the picture has its temptations. If we took it seriously, it would enshrine our familiar three dimensions of space and one of time as the reality to which Hilbert space is a mere approximation, abolishing the unwanted multitudes of extra states that can be derived from Hilbert space.

And my speculation is not quite so wild as it may appear. The concept of subinformation is not new. Quantum has always seemed to imply that the universe somehow "knows" more behind-the-scenes information than can be measured in any one world line. For example, consider a simple quantum entity, a photon that has passed through a polarizing filter set at 38.123456789 degrees to the horizontal. An experimenter can only read one bit of information about the photon's subsequent polarization state. But there is a sense in which the universe seems to know the angle of polarization far more exactly, because the photon is certain to pass through a second filter it encounters later only if that filter is also set at precisely 38.123456789 degrees, and not at any other angle.

(Physicist readers may also recognize a certain relationship between this way of looking at things and the work of Ilya Prigogine. However, I am really reversing Prigogine's argument, which is essentially that in certain contexts, notably the thermodynamics of gases, it is more fundamental to think of matter as a process than as a set of atoms in specific positions, because our ignorance of the atoms' positions is fundamental to the gas having the properties it does. I am suggesting that we should regard Planck-level subinformation as being as real as we normally consider atoms to be, even though we can never read it directly.)

Let us be clear. The picture I am proposing differs from orthodox quantum mechanics; I am replacing the putatively infinite measures of worlds depicted in Figure 12-1c with something more like the picture of 12-1b, in which huge but finite ratios of numbers of different worlds reproduce (to an extremely close approximation) the outcome probabilities predicted by orthodox quantum mechanics. But this is not some wild defiance of Occam's razor. In January 2004, just months before I wrote this chapter, David Deutsch published a thought-provoking paper entitled "Qubit Field Theory"[8] in which he demonstrated that conventional quantum mechanics places no limit on the information that can be described in a limited region of space. Quantum mechanics must be modified to cope with Bekenstein's bound.

I would like to propose a program to see if such a hidden-local-variable multiverse theory is indeed possible, and flesh out its constraints and details. The first stage in such a project might be to write a simple computer algorithm or model that makes use of the following suggestions:

1. It must follow a simple deterministic update rule, with the state of each Planck volume of space (probably represented by a single pixel on the screen) changing each time step in a local way determined only by its own state and that of its immediate neighbors.

2. The updating must give rise to an ever-growing multiplicity of divergent stable patterns that interact significantly with their own "worlds" but little with divergent ones. Different world lines should be made distinguishable to the eye by appropriate use of color and perhaps flashing pixels at different rates.

3. Make it possible for a pattern to give rise to a large number of "daughter" patterns as the result of a single branching event; the relative numbers of the daughter patterns should conform to something analogous to the Born rule.

4. Consider the following mechanism for "condensing" an increasing number of stable patterns from time-symmetric rules. A 1-centimeter cube containing a given number of particles is about 10^{33} Planck lengths on a side. Every second, cosmological expansion increases each side by about 10^{15} Planck lengths, vastly increasing the

amount of information about the particles that we can know from one particular universe viewpoint.

5.   As Penrose has pointed out, space-time should curve differently as perceived in different world-lines as masses move to different positions. The presence of a large amount of nearby mass should make local processes proceed more slowly in a given world-line, because time flows more slowly in a gravity well. Can the model replicate these effects?

Feel free to check my Web site for any progress on this program: http://www.colinbruce.org.

⌒⌒⌒

If we turn out to live in such a comprehensible place as a multiverse of hidden local variables obeying classically deterministic rules—which in my highly personal opinion would be the ultimate extrapolation of the Oxford Interpretation—things will arguably have turned out the best we could have hoped for in all possible worlds. We will inhabit a universe strange enough to fascinate, yet one capable of being visualized with our simple ape brains, run by a clockwork that scientists of the past such as Newton and Laplace would have understood, a universe in which we can hope to play tick-tack-toe with the gods.

We might even be able to explain definitively *why* we find ourselves in such a universe. Just as the optical-fiber computer we met earlier could perform thousands of calculations in parallel using no more hardware than required for a single conventional computer, so a slight difference in the laws of physics could make the difference between a universe that can run only one world line and a multiverse that can run a colossal number. If a multiverse represents a vastly more efficient use of resources, in terms of the number of intelligent species or individual beings it can contain per unit of information processed, then is it not statistically almost inevitable that we find ourselves in such a place?[9]

But now I am treading very close to the line that separates physics from metaphysics, and it is definitely time to bring this book to a close.

THE PRINCIPAL PUZZLES
OF QUANTUM

PPQ 1

Spooky quantum links seem to imply *either* faster-than-light signals *or* that local events do not promptly proceed in an unambiguous way at each end of the link.

PPQ 2

Spooky quantum links seem to imply *either* faster-than-light signals *or* that quantum events are truly random.

PPQ 3

Why does the universe seem to waste such a colossal amount of effort investigating might-have-beens, things that could have happened but didn't?

PPQ 4

Why does reality appear to be the world in a single specific pattern, when the guide waves should be weaving an ever more tangled multiplicity of patterns?

NOTES

Chapter 2

1. By the time Compton did his experiments in 1923, this was the expected result. Einstein won his first Nobel Prize for describing the related photoelectric effect, explaining the way individual electrons are knocked from solid materials by individual photons. Planck performed the first theoretical calculations of photon momentum and energy, although he did not take his hypothetical photons seriously.

2. A more advanced mental picture I like to use, which catches both the surfer and the wave in a single system, is a hoop which, like a Mobius band, has a twist in it. Imagine that the hoop is made of very stretchable material but is resistant to being twisted, so that it stores some elastic energy in the twist. The twist is not uniformly distributed round the hoop. At any given point, the degree of twisting corresponds to the amplitude of the wave. Sooner or later, the ring snaps at some point—most likely somewhere the local twisting is greatest—and untwists itself, reforming as a simple hoop, no longer a Mobius strip, and no longer storing any elastic energy. The hoop-with-twist metaphor works only in two dimensions, but physics-knowledgeable readers can think in terms not of a Mobius strip but a *skyrmion*, a corresponding kind of topological knot in three-dimensional space.

Chapter 3

1.   For example, if the electron detector is a passive loop of wire that has a current induced in it only when a charged particle passes nearby, it still has some effect on its neighborhood at other times, because random thermal motions of electrons in the wire loop will produce a tiny, fluctuating magnetic field.

Chapter 5

1.   Including more-sophisticated experiments involving three particles rather than two, whose results are even harder to quibble with.

2.   Howard, D. 2003. Who invented the Copenhagen Interpretation? A study in mythology. Available at: http://www.nd.edu/~dhoward1/Copenhagen%20Myth%20A.pdf

3.   Bohm, D., and B. J. Hiley. 1993. *The Undivided Universe.* New York: Routledge.

4.   Price, H. 1996. *Time's Arrow and Archimedes' Point: New Directions for the Physics of Time.* Oxford: Oxford University Press. The constraint in our future would probably be different from the known constraint in the past, the pointlike Big Bang. It would be a state of micro order rather than macro order. A visual analogue would be a clump of seaweed at low tide. At the seabed the strands all start at the same point (macro order, the Big Bang); at the surface the strands are spread apart, but wind and buoyancy force them to lie exactly parallel to one another (micro order).

5.   For example, the existence of particle interactions that exhibit what is called CPT violation are a problem for Price's version. This stands for charge-parity-time violation. The particles do not behave in a fully time-symmetric manner.

Chapter 6

1.   There are many possible quibbles with the exact figure. Cosmologists can feel free to add a few orders of magnitude.

2. Joos, E. 1999. Elements of environmental decoherence. In P. Blanchard, D. Giulini, E. Joos, C. Kiefer, and I.-O. Stamatescu (eds). *Decoherence: Theoretical, Experimental, and Conceptual Programs.* Heidelberg, Germany: Springer.

Chapter 8

1. See www.lhup.edu/~dsimanek/fe-scidi.htm.

2. Russell, J. B. 1991. *Inventing the Flat Earth: Columbus and Modern Historians.* New York: Praeger.

3. Gribbin, J. 2002. *Science: A History 1543-2001.* London: Penguin Books, pp. 421-424.

4. Hoping that you can ignore the effects of far-off things because their influence is relatively small is not necessarily justified. For example, the inverse-square law tells you that many forces diminish by a factor of four for a doubling of distance, but a doubling of distance implies that you must then take into account the effects of objects in an eightfold greater volume of space. If the action is instantaneous, you can get the kind of self-reinforcing interactions that we nowadays call positive feedback. There is never a guarantee that the universe will be comprehensible, but a universe incorporating instantaneous long-range interactions is likely to be almost impossible to get to grips with.

5. Turnbull, C. M. 1961. *The Forest People.* New York: Simon & Schuster, quoted in R. D. Gross. 1987. *Psychology, the Science of Mind and Behaviour.* London: Hodder & Stoughton, p. 129. I am indebted to Claire Chambers for tracking down the source of this story.

Chapter 9

1. Deutsch, D., and P. Hayden. 2000. Information flow in entangled quantum systems. Centre for Quantum Computation, The Clarendon Laboratory, University of Oxford. *Proceedings of the Royal Society of London, Ser. A* 456:1759-1774.

2. Tegmark, M. 2003. *Scientific American,* May. An expanded version appears in the online physics archive http://www.arxiv.org.

Chapter 10

1. Kwiat, Zeiliger et al., High-efficiency quantum interrogation measurements via the quantum Zeno effect, arXiv:quant-ph/9909083 v1 27 Sep 1999.

2. Paul, H., and M. Pavicic. 1997. Nonclassical interaction-free detection of objects in a monolithic total-internal-reflection resonator. *Journal of the Optical Society of America* B 14:1273-1277.

3. Kent, A., and D. Wallace. Quantum interrogation and the safer X-ray. Quantum Physics, abstract quant-ph/0102118 v1.

Chapter 11

1. Feynman, R. 1982. Simulating physics with computers. *International Journal of Theoretical Physics* 21 (6/7):467-488.

2. Deutsch, D. 1985. Quantum theory, the Church-Turing principle, and the universal quantum computer. *Proceedings of the Royal Society of London, Ser.* A 400:97-117.

3. http://www.chem.ox.ac.uk/curecancer.html.

4. http://www.climateprediction.net/index.php.

5. Other discoveries such as Grover's search algorithm still require further work before they can truly be said to do anything "useful."

Chapter 12

1. Vaidman, L. 2002. "Many-worlds interpretation of quantum mechanics." In E. N. Zalta (ed.), *The Stanford Encyclopedia of Philosophy* (Summer ed.), Available at: http://plato.stanford.edu/archives/sum2002/entries/qm-manyworlds/.

2. Deutsch, D. 1985. Quantum theory, the Church-Turing principle and the universal quantum computer. *Proceedings of the Royal Society of London, Ser.* A 400:97-117.

3. Deutsch, D., and P. Hayden. 2000. Information flow in entangled quantum systems. *Proceedings of the Royal Society of London, Ser.* A 456:1759-1774. Available at: http://xxx.lanl.gov/abs/quant-ph/9906007.

4. Deutsch, D. 2004. Qubit field theory, January. Available at: http://arxiv.org/ftp/quant-ph/papers/0401/0401024.pdf.

5. Wallace, D. 2003. Everettian rationality: Defending Deutsch's approach to probability in the Everett interpretation. Quantum Physics, abstract quant-ph/0303050 revised March 11.

6. Another notable British example is Jim Lovelock, famous for his discovery that an ecosystem is unstable until it becomes limited by the chemical resources available to it. A consequence is that the atmosphere of any life-bearing planet should deviate from chemical equilibrium, so planets with ecosystems should be detectable from afar by looking for excesses of such gases as ozone and methane. Lovelock's concept of "Gaia" to describe the Earth's dynamic equilibrium made him a darling of the early eco-movement. His income from ingenious patents made an academic post unnecessary.

7. Barbour's more technical work, which is too complex for us to go into here, essentially concerns the problem of how we can specify the state of the whole universe at a particular instant when, due to relativity, different observers do not agree on what constitutes a simultaneous instant.

8. Gell-Mann, M. and J. B. Hartle "Strong Decoherence" In D.-H. Feng and B.-L. Hu (eds). *Proceedings of the 4th Drexel Conference on Quantum Non-Integrability: The Quantum-Classical Correspondence.* Hong Kong: International Press of Boston. arXiv:gr-qc/9509054 v4 23 (Nov).

Chapter 13

1. http://www.hep.upenn.edu/~max/index.html.

2. Lewis, D. 2004. How many lives has Schrödinger's cat? *Australasian Journal of Philosophy* 82:3-22.

3. In the full and nightmarish version of Jonathan Swift's *Gulliver's Travels*, which is not normally given to children to read, the plight of these immortal but terribly enfeebled and senile persons is graphically described.

4. Sebastian Sequoia-Jones, conversation with the author, March 2004.

5. David Wallace, conversation with the author, November 2003.

6. Piccione, M., and A. Rubinstein. 1997. On the interpretation of decision problems with imperfect recall. *Games and Economic Behavior* 20:3-24.

7. Vaidman, L. 2001. Probability and the MWI. In A. Khrennikov (ed.), *Quantum Theory: Reconsideration of Foundations*. Vaxjo, Sweden: Vaxjo University Press, pp. 407-422.

Chapter 14

1. Penrose, R. 1989. *The Emperor's New Mind.* New York: Oxford University Press.

2. Marshall W., C. Simon, R. Penrose, and D. Bouwmeester. 2002. Towards quantum superpositions of a mirror, Quantum Physics, abstract quant-ph/0210001, revised September 30.

Chapter 15

1. To a many-worlder like myself, this "tip-of-the-iceberg-effect," the discrepancy between the large amount of information that the universe needs to know about the particle (the exact angle of its spin) to make it behave appropriately, and the single bit that can be read out in any given "world," can be seen as further evidence for the existence of the multiverse.

2. For further discussion of this lattice-based approach, including a description of Planck lengths and the holographic principle, see Chapter 16.

3. Normal tolerances in the process of printing, folding, and binding these book pages may result in an inexact superimposition of Figures 15-2 and 15-3, thus preventing the stated effect from occurring. To observe it, the reader may photocopy both figures and hold them back to back against a strong light, adjusting the superimposition carefully until the effect appears.

4. Many readers will have realized that this is just a variant of the one-time-pad still used for sending secure messages today.

5. Zeilinger has attempted to develop his system using an alternative measure of information to that given by conventional Shannon

information theory. He believes that this approach is justified because the classical "ignorance" interpretation of probability described in Chapter 5 is not adequate in a quantum context. The validity of this claim is vigorously disputed by many theorists.

Chapter 16

1.   Shahriar, A. 2004. "Quantum Rebel." *New Scientist,* July 24, 2004, p. 30.

2.   Plaga, R. 1997. "Proposal for an experimental test of the many-worlds interpretation of quantum mechanics" Found.Phys. 27 559. http://xxx.lanl.gov/PS_cache/quant-ph/pdf/9510/9510007.pdf.

3.   Vaidman, L. 1996. On schizophrenic experiences of the neutron, Quantum Pysics, abstract quant-ph/9609006, revised September 7.

4.   Tegmark, M. Does the universe in fact contain almost no information? *Foundations of Physics Letters* 9:25-42.

5.   This statement of course needs qualifications. For example, if the galaxies involved contain not just pointlike stars but clouds of gas and dust, as most or all galaxies do, there will be significant interactions between those entities that can trigger bursts of star formation and other effects. But the point I am trying to make is that perfectly classical physics can include things that share the same volume of space, but interact relatively little with one another.

6.   Feynman, R. 1994. *The Character of Physical Law.* Cambridge, Mass.: Modern Library.

7.   Bekenstein, J. D. 1973. Black holes and entropy. *Physics Review* D7:2333-2346.

8.   Deutsch, D. 2004. Qubit field theory, January. Available at http://arxiv.org/ftp/quant-ph/papers/0401/0401024.pdf.

9. We could take this anthropic argument a step further. One of Oxford's most famous authors, C.S. Lewis, speculated that the vastness of cosmic distances might represent "God's quarantine regulations," ensuring that an imperfect species such as our own could not extend its influence to other worlds. We now know that his hope was false: Travel over interplanetary and even interstellar distances is defi-

nitely possible for a technologically advanced species. Indeed, astronomers wondering how many intelligent species our universe may contain have seriously considered what is called the Queen Bee hypothesis. There is normally only one queen in a hive of bees, because the first new queen to be born promptly stings any potential rivals to death in their larval cells. An intelligent species that develops interstellar travel might well use its power similarly to ensure that it would never have any dangerous competitors. In that case, there will usually be only one intelligent species per universe.

The same logic would apply to the multiverse as a whole—if there was any way at all in which creatures occupying one small slice of it could reach out to affect other "parallel worlds." For a multiverse to support a huge number of species, we do not need merely laws of physics that efficiently support multiple processes. They must embody a very special combination of properties, for they must *also* in some subtle way make it not just technologically difficult, but fundamentally impossible, for a being, however intelligent, to systematically affect world lines far removed from its own. That is exactly what we are currently discovering.

INDEX